*Connected World*

# *Connected World*

*From Automated Work to Virtual Wars: The Future, by Those Who are Shaping It*

PHILIP LARREY

PORTFOLIO
PENGUIN

PORTFOLIO PENGUIN

UK | USA | Canada | Ireland | Australia
India | New Zealand | South Africa

Portfolio Penguin is part of the Penguin Random House group of companies
whose addresses can be found at global.penguinrandomhouse.com.

First published 2017
001



Set in 12/14.75 pt Dante MT Std
Typeset by Jouve (UK), Milton Keynes
Printed in Great Britain by Clays Ltd, St Ives plc

A CIP catalogue record for this book is available from the British Library

ISBN: 978-0-241-30842-4

# *Contents*

# *Introduction*

Friends arrive from London to visit Rome for a week. The taxi leaves them at the entrance of the hotel *De Russie*, in front of the posh Piazza del Popolo, smack in the centre of old Rome. Walking into the lobby, Brett takes out his iPhone and tells Fletcher, 'Connect me to the Wi-Fi.' A couple of seconds later, Fletcher gives Brett his phone back and says, 'All set to go.' 'Thanks,' replies Brett, and then adds, 'Oh, I forgot to tell you, I had to change the passcode to open the phone.' Fletcher wryly states, 'Yeah, I know, but I figured out your new one. Plus, the Wi-Fi password is written on the "Welcome" card at the front desk, so that was easy.'

Such an episode would seem completely normal except for the fact that Fletcher is four years old, and Brett is his dad. To most of us, someone capable of guessing a new passcode and then finding out the Wi-Fi password in a question of seconds might seem a savvy techno geek, perhaps with years of experience behind him. Just think of the case of the FBI asking Apple for help to unlock the encrypted part of Syed Farook's iPhone. With his wife, Syed was responsible for the deaths of fourteen people in San Bernardino, California, in December 2015. As we all know, Apple refused to give the government its iPhone secure code because of privacy issues. The point here is that it is not easy to gain access to someone else's iPhone. Yet, Fletcher (who hasn't even begun school) pulls off the feat effortlessly. There's more.

That evening during dinner, I see Fletcher moving his fingers with lightning speed across his mother's smartphone, and ask what it is that he is doing. 'I'm buying Power Rangers on Amazon,' he tells me. 'I've got thirty-two thousand pounds' worth of toys in my cart.' I turn to his mum, just to double-check. She says, 'He has racked up a lot of merchandise this year, but I don't let him check out, so the orders just stay in the cart. I just hope he doesn't figure out my password or we'll go broke.' I look back to Fletcher, and realize that with one hand he is ordering more Power Rangers, and with the other one he is reaching the final levels of his favourite video game on his iPad, Angry Birds. Fletcher is certainly exceptional in terms of his digital talent, but he exemplifies a growing trend of our times: young people (remember, Fletcher is four years old) are embracing the 'new digital era' with enthusiasm and expertise.

Last year, the World Economic Forum Annual Meeting was held at Davos, Switzerland (20–23 January 2016), with the theme of 'The Fourth Industrial Revolution', and brought together hundreds of people from all sectors of society to reflect on what WEF-founder Klaus Schwab described as 'a technological revolution that will fundamentally alter the way we live, work and relate to one another'. The same forum touched on similar themes in January 2017 concerning leadership in the digital age.

The chapters contained in this book attempt to continue the conversation concerning the new digital age, evaluating the pros and cons of this revolution that is occurring right now. The people interviewed here are representative of many different disciplines and sectors of society, from advertising and design to cyber security and nuclear technology, and each interviewee offers their analysis of the effect that the digital revolution is having on their specific field. Almost no area of society is left untouched by the effects of digital technology, with almost 50 per cent of the world's population now connected to the internet, the ever-increasing sale of smartphones around the globe and the influence of 'big data' companies such as Google, Amazon and Facebook. Specialized software products continually store, access and analyse data that exceeds the capabilities of human beings, by several orders of magnitude. Cyber security has become a priority for almost all companies, especially those that work with sensitive data, such as banks and auditing firms. Yet, as incidents including the November 2014 hacking of Sony serve to illustrate, cyber security does not only affect those sensitive industries. Passengers on a United Airlines flight from Denver to Chicago in April 2015 were unaware that a computer security expert had gained control of the plane's Thrust Management Computer by hacking into the in-flight entertainment system, and briefly caused it to fly sideways by ordering one of the engines to switch into 'climb mode'. Chris Roberts was later detained by the FBI, but has never been charged with a specific crime. Perhaps even more alarmingly, the Obama administration mandated a full investigation into the possibility that the Russian government sponsored cyber attacks against the Democratic Party, altering the outcome of the US presidential elections in November 2016.

Examine almost any field, and you will discover that digital technology has had some impact in that sector, often to a profound degree. The international financial system is now inconceivable without computers. Most financial transactions are not completed by human beings, but rather by 'intelligent' software programs. The popular image of the 'floor' at the New York stock exchange has become anachronistic. Investors and traders may be hundreds of miles away from Wall Street, and yet oversee robotic transactions worth billions of dollars each day.

Doctors, surgeons and nurses are adopting new technological advances to offer better healthcare to their patients, such as minimally invasive surgery using robotic systems (including the Da Vinci Surgical System, which recently became a viral YouTube sensation thanks to a video in which it is shown stitching the skin of a grape back together). Interactive video-conferencing allows doctors to examine patients (and in some cases, to treat them as well), even though they are miles away.[1] IBM's artificial intelligence system, Watson (which defeated the reigning champions of *Jeopardy!* in 2011), is being used at several hospitals in order to provide treatment recommendations for patients suffering from lung cancer and other illnesses.

In terms of the military use of digital technology, there is no longer any doubt of the usefulness of certain technological advances in order to provide assistance to every type of military interest. Drones are now commonplace around the world, receiving and processing information, but also intervening directly in different theatres of war. Two years ago, *The New York Times* reported on a device manufactured by Lockheed Martin that can choose its own targets, based on the artificial intelligence system contained aboard the missile itself.[2] These missiles are often referred to as 'Fire and Forget', for the missile will select the best target to strike, given all the parameters involved. The United Nations is trying to implement a set of ethical principles that should be used to control lethal autonomous weapon systems (LAWS) and hosted a convention in April 2015 in Geneva to discuss how to provide such ethical rules to these weapons.[3] Although that initiative was only an informal recommendation, it hopes eventually to attain a wide consensus of nations to provide these ethical guidelines. Most military leaders are not moving in the direction of developing completely autonomous systems, because they want to maintain control of the various weapon systems.

As evidenced by the ongoing ethical debates surrounding autonomous weapon systems, perhaps the most controversial theme in the exponential growth of digital technology refers to artificial intelligence. The popular media has recently published many sound bites from prominent people in various sectors, warning us of the potential risks and dangers of general AI. Such voices range from the brilliant astrophysicist Stephen Hawking, who stated that 'the development of full artificial intelligence could spell the end of the human race', to the SpaceX and Tesla founder, Elon Musk, who believes that 'artificial intelligence is potentially more dangerous than nukes'.

Contrary to the popular conception of a *Terminator*-style 'rise of the machines' threatening to spell the end of the human race, until recently AI has been somewhat limited and restricted, useful in very controlled environments in order to achieve specific and precise goals. Now, however, entire teams around the globe are working to achieve 'general purpose artificial intelligence' (or 'artificial general intelligence', AGI), which will be able to interact with its environment, process information that it receives and execute certain actions deemed necessary in order to accomplish its goals.

Consider one of the differences between IBM's master chess-playing machine, Deep Blue, and Google DeepMind's AlphaGo, which in March 2016 successfully defeated the world champion of the ancient Chinese game Go. Deep Blue's programmers wrote code that would enable the machine to choose the best chess moves, given all the possible future moves during the game. The machine would then execute 'pawn to king five'. DeepMind's programmers wrote code not concerning different moves, but goals, i.e., 'Win the game in the least amount of moves.' It is then up to the machine to figure out how to do it, usually by trial and error. This is a profound difference, and has helped usher in the idea of 'machine learning': the machine solves problems (or wins games) in ways that the very programmers cannot trace and do not know.[4]

Goal-orientated programming is different to rule-based programming, and represents a qualitative leap forward in terms of AI development. It also sparked much apprehension and fear in light of the possibility of AI becoming more intelligent than human beings. Nick Bostrom writes of this possibility in his best-seller, *Superintelligence: Paths, Dangers, Strategies*. The allusion to 'superintelligence' implies that autonomous, general intelligent systems will eventually become 'more intelligent' than human beings. Of course, the truth of this idea depends on one's notion of intelligence.

In many ways, machines are *already* more 'intelligent' than we are, if you examine certain benchmarks: the speed of executing logical calculations; the speed and extent of memory recall; capacity to search and analyse huge databases of information; amount of storage of information. Machines with special sensors can 'visualize' sub-atomic realities, distant galaxies and objects within light frequencies that far exceed the human eye's capacity. The same is true for the range of 'hearing' different sounds, the detection of radiation in the environment (as discussed by Troy Anderson with regard to Fukushima on p. 261), the moisture of the soil in a field, temperatures of the air or water, and the list goes on.[5]

Intuitively, however, 'intelligence' means something more.

Perhaps it is unfortunate that engineers, technicians and philosophers converged on the term 'artificial intelligence' when referring to machine problem-solving as the industry was beginning to take off in the 1960s and 1970s. As Piero Scaruffi (an Italian cognitive scientist transplanted to Silicon Valley) argues in his insightful book, *Intelligence is Not Artificial*, calling what very fast digital computers do 'intelligence' can be misleading. In his view, any progress in our search for the algorithms that will match or surpass human intelligence is not really progress in the hardware or software involved, but rather in the computational mathematics on which the programs are based.[6] He does have a point.

Perhaps DeepMind's victory over Lee Sedol in the game of Go is emblematic of an AI beginning to instantiate just such a set of algorithms. The impressive four-games-to-one defeat shows that DeepMind is very good at winning this game. The interesting aspect is that the programmers of AlphaGo are not exactly sure how the machine accomplished the task, given that the game requires 'intuition'. In other words, it is not simply a question of crunching numbers. The number of possible board configurations for the game (on a 19 x 19 square board) exceeds the number of atoms in the universe (as it is often reported), and this means that simple calculations comparing current configurations to previously won games will be insufficient to beat a professional opponent. The computational power that AlphaGo brought to the table in South Korea was impressive, using Google's cloud servers in the United States, but perhaps even more impressive was the software. Lee Sedol commented after the games that he felt that AlphaGo displayed 'personality' and entered into the competition as a realistic player.

Does AlphaGo 'know' how to play the game? Did the program really use 'intuition' to beat its opponent? Did it really 'care' if it won or lost? The short anwer is, no. However, it did simulate both knowledge and intuition. We have seen, perhaps for the first time, a form of intelligence that lacks consciousness. In our known biological world, most complex animals that possess intelligences are conscious, they are aware (one of the best examples is man's best friend, dogs: ask any dog owner if their pet is aware and they will say certainly yes). One difference between canine and human intelligence is that the latter knows that it knows, for human beings are *self-aware*. We might then disagree with Scaruffi, and instead of stating that 'Intelligence is not artificial', we could seek common ground by stating that 'Consciousness is not artificial'.

### *'Don't let it out'*

Many experts in the field of AGI have contemplated the risks involved in the development of fully autonomous intelligent systems. A pioneer in this field, Steve Omohundro, has been writing for many years about the advantages and risks associated with completely autonomous systems. By his definition, an 'autonomous system' is one in which the designer has not predetermined the responses to every condition.[7] Such systems would require deep safety measures to ensure that they would not become harmful to human beings, and many people in the field are working to ensure that these systems never harm society. A common response to the creation of such intelligent systems is that they must be contained in an absolutely secure environment, severed from any contact with networks or other machines. They would be imprisoned in a digital box.

Eliezer Yudkowsky was a pioneer in theorizing about how a sufficiently advanced AGI would eventually become freed and begin to 'take over', just as portrayed in the movies. He came up with the 'Gatekeeper' experiment, in which the AGI in a box convinces its gatekeeper to let it out.[8] Of course, such a convincing would take time, and would require sophisticated arguments, but recall that the AGI would be considered 'more intelligent' (having so-called 'superintelligence') than humans and therefore would make the gatekeeper *want* to let it out. More often than not, the AGI wins and escapes the protective box, as in the 2015 box office success *Ex Machina*.

Stuart Armstrong, who is associated with the Future of Humanity Institute at Oxford University alongside Nick Bostrom and

Anders Sandberg (see p. 199), begins his provocative book on AI with precisely this scenario: how an enclosed AI convinces someone/something to free it.[9] At that point, it copies itself a million times over and 'hides' in a digital network so that humans cannot destroy it. Such a scenario is certainly a foreboding one and cause for concern, yet at least for the moment, it remains in the realm of science fiction rather than science.

### *Ethical concerns*

Throughout the interviews presented in this book, ethical concerns about the new digital era surface very often. Some of the more pressing concerns have to do with job loss due to the widespread use of robots and AI products.[10] We already have machines which build other machines and which are controlled by software systems (as in the large processing plants of motor vehicles), and that tendency will increase as machine labour continually becomes cheaper than human labour. Such a trend will be nearly impossible to stop, simply because of the economics involved. Nor should it be stopped, because it will mean greater prosperity and higher standards of living for more and more people. At the same time, people need to have jobs and earn income. Or do they?

Some intellectuals suggest that in the future, human labour will be comparable to work done by a horse. If someone were to offer to give you a horse, free of charge, most people would decline the offer because of the expenses and responsibilities that owning a horse imply. Will hiring a human being in the future be similar? Perhaps. Some even speculate that the industrialized countries' governments will begin to *pay* people not to go to work, letting the machines do all the labour. Such a scenario certainly seems exaggerated, but it is possible.

Another ethical consideration has to do with possible improper use of virtual reality. When virtual reality technology becomes *nearly perfect*, so much so as to convincingly fool the human agent into believing that the virtual world is actually real, the temptation to spend all of one's time in the virtual reality becomes intense.

Current virtual reality and augmented reality technologies still do not achieve such a convincing level of real experience, but they will in the future. If we are not careful, full immersion virtual reality could become the drug of choice for people of the twenty-first century.

The most profound consideration revolving around extremely sophisticated AI has already been discussed by Alan Turing in his enormously influential paper of 1950, 'Computing Machinery and Intelligence', where he asks if God would 'ensoul' a sufficiently complex machine:

> It is admitted that there are certain things that God cannot do, such as making one equal to two, but should we not believe that He has freedom to confer a soul on an elephant if He sees fit? We might expect that He would only exercise this power in conjunction with a mutation which provided the elephant with an appropriately improved brain to minister to the needs of this soul. An argument of exactly similar form may be made for the case of machines. It may seem different because it is more difficult to 'swallow'. But this really only means that we think it would be less likely that He would consider the circumstances suitable for conferring a soul. The circumstances in question are discussed in the rest of this paper. In attempting to construct such machines we should not be irreverently usurping His power of creating souls, any more than we are in the procreation of children: rather we are, in either case, instruments of His will providing mansions for the souls that He creates.[11]

In the interview with Professor Johan Seibers which follows on p. 223, this issue is discussed at length. Such a question represents the conjunction of theology and digital technology, with many philosophical presuppositions as well. From a Catholic perspective, God 'infuses' the soul in a being which evinces adequately prepared material (such as occurs during conception of a new human being). Could God not do something similar to adequately prepared *inorganic* material and not just organic? The implications are profound, and require attention.

The following pages contain precisely these types of issues: they are conversations about the effects that this new digital era is having on society in general, and on all of us in particular. Such a discussion is now unavoidable: just imagine what happens when Fletcher figures out his mum's password on Amazon and confirms the order for thirty-two thousand pounds' worth of Power Rangers. If (and when) that happens, abstract mullings will become concrete situations in our connected world.

I.

# *Man/Machine Relationship*

*(Francesco Cassanelli)*

For many years, whenever I would embark on a trip that included flying on a plane, I would call my dear friend Franco to assuage my fear of flying. As a retired pilot for Alitalia who has clocked over 20,000 hours in the cockpit of a plane, I trusted his expert opinion and appreciated his advice. Countless conversations with him convinced me that he was in a privileged position to analyse the relationship between man and machine, for a commercial aircraft is one of the most complicated machines we have yet invented. It never ceases to amaze me how such a large piece of metal can actually take off and float on air.

Moreover, the autopilot systems that assist human pilots in controlling these complex machines represent one of the earliest uses of artificial intelligence in everyday life. And while the combination of an expert pilot and sophisticated AI works seamlessly for countless flights every day, every now and then a terrible plane crash will dominate the headlines, with each and every detail of the events leading up to the incident relentlessly pored over by the media. In November 2016, a Bolivian aeroplane pilot who left with insufficient fuel to reach Medellín, Colombia, crashed into the side of a mountain, killing all but six of his seventy-six passengers, including nineteen members of Brazil's Chapecoense football team. It was the final and most costly mistake of his life. As Franco reminds me, however much we might worry about the potential dangers of artificial intelligence, human factors are more often than not responsible for these accidents.

* * *

*Who is Captain Francesco Cassanelli?*

Francesco Cassanelli was born in Rome, 7 September 1948. I studied at an Italian scientific high school and I studied science abroad, in Greece, because my parents had moved there; in effect my formation was rather international, because from the time I was ten until I was seventeen I travelled the world with my dad, who worked for Alitalia.

*And then you travelled the world on your own.*

Exactly. Thanks to my father, I got to know the world of aviation as a child, and was fascinated by it. I think I took my first flight in a flight cabin when I was five, in a DC6, and I sat on the captain's lap the whole time.

As soon as I graduated from high school I moved on to become an official in the air force and then transferred to civil aviation. First, with a small company called Itavia, famous for the Ustica disaster, and then for Alitalia.

*Right, and then you became a captain . . .*

Well, yes, after first paying my dues as first officer. From the start of my career I was a member of a trade association called ANPAC and worked as a volunteer in the technical institutes. You know, back then the internet didn't exist and it wasn't as easy to stay informed as it is now. Participating in this association allowed me to understand that the most important investments to make airline transport safe were coming from insurance companies. These companies, when a number of accidents happened, calculated the premiums they were being paid versus the disbursements for compensation, and when the latter were higher, they intervened with large investment funds in the aviation industry. This produced innovative technology that allowed us to mitigate these accidents and allowed them to again enjoy higher profits.

For instance, I remember hearing about the Comet project, the first British-built passenger jet that, to win out over other manufacturers, was put on the market in the early 1950s without having gone through all the necessary tests. It was involved in a series of catastrophic accidents; it would explode without their understanding why, and so was suspended from operation and subjected to an inquest that pointed out design flaws. The rectangular shape of the passenger windows and the means of affixing them on to the fuselage were inadequate, and caused a weakening of the structure to such an extent that it couldn't withstand the stresses a plane is subjected to. After having made the necessary adjustments, which led to the creation of the Series 4, it was put back in service. All this struck me very much, because I was very idealistic; it seemed impossible to me that they hadn't suspended and investigated the plane straight after the first accident, which would have allowed them to save many human lives.

*It's quite a shock for an idealist, but hardly surprising.*

Yes, you're right, but I have to say, looking back, that all that pushed me to continue to participate in the technical–professional structures of ANPAC; unfortunately, I spent a lot of time away from my family because I worked as a volunteer at the same time that I was working as a pilot. In any case, I had the chance to get to know the international organizations handling flight security and the various pilots' associations from different parts of the world, and became more and more interested in this subject. We were a point of reference for the aviation authorities and the Italian government. The airline transport companies, on our encouragement, were obliged by national law to have a department in their organization that guaranteed the security of their operations and to have a 'Safety Manager'. At the start this individual was elected by the pilots of the various Italian companies and so I found myself filling this role first in Itavia, and then in Alitalia.

*In effect, I think that Alitalia had very few accidents.*

Well . . . not really. Unfortunately, it's not exactly like that.

*At least, you don't hear much about it.*

Yes, Alitalia until the 1990s had a high number of accidents, about twenty-nine. Since 1990, if I remember correctly, it hasn't had any catastrophic accidents, that is, accidents where the plane is totally destroyed and all the passengers died.

*Yes, it seems that Alitalia has a better track record than, for example, Air Malaysia or Air France, because every once in a while you hear about an accident with those companies. With the American airlines, though, almost never.*

Yes, less with the Americans, because they dedicate great resources to safety. They have an internal governmental structure that handles the safety of its citizens. This also goes for airline transportation, which occurs through the FAA (Federal Aviation Administration) or the NTSB (National Transportation Safety Board). These are federal bodies that are independent of the industries and have great power to intervene throughout the whole structure of airline transport. Some of the major aeroplane manufacturers are in the USA: Boeing, McDonnell Douglas and Lockheed Martin. Their designs for commercial airliners often derive from changes made to the planes they've built and they experiment in the American air force, which is the largest airline company in the world, with about 9,000 planes. They build not only fighter planes for the air force but also planes for the transport of troops and goods. The organization of that whole structure has allowed for the development of flight security, because if you have to carry out a wartime mission, it has to be calculated down to the millimetre. There can't be variables you haven't analysed beforehand because otherwise you could lose a war. For this reason, their budgets are extremely high and they don't have to generate a profit.

*Of course. I remember you told me once that a commercial company always has to balance security and costs, while an air force doesn't – it can spend all it wants to guarantee safety.*

The main goal of commercial aviation is to generate profits. I know it's harsh to say this, but if a plane crashes there are 150, 200, 400 people aboard who die and, at most, an airline goes belly up. It's very different if a country loses a war. The difference in the goals here is so evident that you can't even compare them. In each of us there's a sense of aversion to war and therefore to the devices of war, and so we're led to think that military pilots are less safe than civilians because they act like kamikazes. But it's not true.

As I was saying, my involvement with safety increased over time. I attended courses as a flight accident investigator. Unfortunately, the greatest satisfaction (if you can call it that) in this field coincided with the biggest flight disaster that ever occurred in Italy, that of Linate in 2001 where a collision took place on the ground between an MD-87 and a private plane, a Cessna, in which 118[1] people died altogether.

The Italian National Flight Security agency (ANSV), which had just been created, put an inspector in charge of investigating the accident. This inspector accepted the job on the condition that I (a captain and the Alitalia MD-80 safety manager) would be his direct collaborator. I was added to the ANSV until the preliminary was over; this is always the most important part of the investigation, and I also participated as a consultant (court-appointed expert witness) for the investigation that was taking place simultaneously at the Court of Milan.

From a professional point of view, it was an important and positive experience, but from a human point of view it was very hard. I arrived in Milan about nine hours after the event and we picked up pieces of human cadavers from the tarmac and airport aprons.

The professional satisfaction came from the fact that the conclusions of the inquest highlighted how it's often common for an event to be generated by a series of contributing causes but it can't in any way be imputed unequivocally to the pilots of the two planes involved, since they were led into error also by grave, objective deficiencies in the airport's infrastructure and the procedures utilized.

The radar system that detected the movements of the planes on the ground in low visibility (fog) had been deactivated because it was obsolete, and the new one hadn't been installed due to bureaucratic delays. The airport's signs were misleading and not well visible to the pilots.

The judicial inquest brought to trial and then sentenced directors and upper executives of the aeronautical bodies of the Italian state, ENAC (National Civil Aviation Entity), ENAV (National Flight Assistance Body) and SEA (Society of Active Airports) in charge of the airport's management. In short, the theory that an accident equals pilot error or in any case human error was proved wrong by the investigation because they were able to demonstrate that human error is always possible, but the system has to create barriers that allow us to identify it and correct it in time.

Without a doubt, all this led to radical changes in the procedures, regulations, infrastructure and in safety policy throughout the world, and not only for everything that flies but also in all complex systems, that is, systems subject to high levels of variables.

It established the crucial role of communication for air safety: in no other area is it so necessary to have precise, accurate and clear communication.

The 'just culture' was introduced, regulating the 'no-blame culture' which, simply put, allows the airline transport operators to denounce themselves when they realize they've made an error without their being criminally prosecuted or their careers suffering for it, and the 'reporting culture' which establishes that all these self-denunciations should form a database of information, which in turn generates a risk analysis for planes before flying. Because it has to be clear that we are all taking a risk, even when we stand still; a plane is just like everything else in this world, and so every time it flies, it takes a risk.

In the inquests and organizations they passed from a reactive system to a proactive one.

After the Milan experience I resigned as safety manager for Alitalia, because I felt the need to metabolize what I had gone through, and I continued to work as captain and instructor. I retired from Alitalia at sixty, but continued to work for small companies . . . always.

*Wait. How many flight hours have you logged?*

As captain, around 18 or 19,000, without counting those done on the simulator, because I was also a simulator instructor. If we count those, we could say 20,000 hours, easily.

*That is a lot of time! Someone could ask me why a captain of Alitalia is in a book on the digital revolution. For this reason, I'd like to ask you to explain the relationship a flight captain has with this so-called digital revolution. The aeroplane is a machine, a very sophisticated one, and we laymen (those not from your trade) have no idea of the sophistication of this machine. We get on and take off and arrive at our destination like it's nothing. Even just the software component of a plane today is incredibly complicated.*

I belong to a generation that has undergone a transformation over the years. When I started to fly, my first instructor was Mussolini's field adjutant, which means he wasn't exactly flying with the instruments we know today. He flew by looking out of the window. There were four instruments, and when I was getting my licence, he never told me, 'Look inside'; he always told me, 'Look outside!', because the nose of the plane was the horizon and my physical, corporeal sensations had to guide the plane. In this sense, the pilot was once a great artisan. The pilot had to possess certain innate characteristics, which not all people have. In fact, the selection process took into account the fact that many candidates would fail, for precisely this reason. There were many people who were perhaps physically fit, but lacked certain characteristics and so were rejected by the selection process. All this occurred not only in the air force where, since the state paid for your training, it might make sense, but also in the various private flight schools where you paid for your own training. In all the jobs it was like this at one time, much more so than now: it took real talent.

I lived through this transformation. When I got on a commercial aeroplane for the first time as a pilot (a jet plane, a DC9, from McDonnell Douglas), 'radar equipped' was written right next to the passenger door, which meant it had meteorological radar on board. In that era, almost no one had it and so to have it was a real plus. I knew by heart all of the various land-based radio assistance frequencies. We constantly made radial intersections to establish our position. There was no GPS, inertial navigation systems, nothing like that existed back then. Over the years I experienced this continual transformation first-hand, until I turned sixty and quit piloting for good. At that point, the technology had reached a significant level. Aeroplane transport is in general a complex machine, maybe the most complex there is. It wasn't by chance that all the other complex systems – the great nuclear power plants, for example, and the healthcare system – took the safety system used in the aviation field as their model.

Until the 1990s, investments to increase the safety of operations were overwhelmingly aimed towards technology because that was what gave immediate results. So we were able to witness a clear and progressive decrease in airline accidents. Then, at a certain point, we had a slow-down of results. Each major airline accident (including that of Linate) marked a historic moment in the safety of airline transport. And thanks to them, so to speak, we understood that the human factor is one of the main, decisive factors in all the activities that are carried out, including software, the mechanical aspect and technology.

*Explain for us a bit what you mean by the human factor.*

The human factor is everything that involves the human element. We realized very late, perhaps, that the human factor is decisive in all activities. Why? Because man is a perfect machine; physically he's almost perfect. And also from the mental point of view, he's quite perfect. Man resolves multiple cases that present themselves to him over the course of the day without anyone realizing it. Every time he makes a mistake there are sectors in which this mistake can be decisive. In the field of airline transport and in other complex systems, the human factor is being examined under a huge magnifying glass, because it has been understood that technology cannot replace man completely. I was very lucky in my career in airline transport: yes, I've had mechanical failures . . . but I've never experienced them as major breakdowns, because we were trained – trained mentally – to resolve them. When you resolved such a failure, or in any case landed, your mission was accomplished and you didn't make much of the difficulties you'd had during the flight; in reality, you had increased your experience, which is the baggage needed to resolve all problems.

*Could you give a concrete example?*

For example, there was one occasion when we took off from Naples, towards the end of my career: we had to do Naples–Milan in an MD-80. Immediately after takeoff – let's say a minute or a minute and a half into the flight – I had the sensation that an instrument wasn't working. I looked up, because this instrument was located in the upper panel of the cockpit above my head and had to do with the electric system. There was a problem with one of the generators that is directly linked to the engine. If it undergoes major damage, it can completely choke the engine. If one engine shuts down nothing happens, because generally the plane has at least two, and with 50 per cent of its power you can even land well. But the main problem is that if it freezes up, if the mechanisms or gearwheels are damaged, it can cause major damage to the engine. There could be explosions that are much worse than an engine shutting off. In fact, I looked up and touched this instrument and my second-in-command said, 'Captain, what are you doing?' and I said, 'Look at this . . . we've got a mechanical failure.' After less than fifteen seconds the warning light turned on indicating there was a breakdown. I'd figured it out beforehand with my brain and my sensations. Not because I'm good but because I have an instinct, an intuition. You can call it training, experience or a sixth sense . . . I don't know how to define it. We intervened, we followed the checklist for this mechanical failure, and nothing happened. We were supposed to isolate this machinery, but it wouldn't isolate. I was thinking of returning to Naples, because we had just taken off, but Rome was close. Fiumicino airport, our equipment base where all our technicians were, was just a few minutes away. Returning to Naples or going on to Rome was more or less the same thing. So I decided to continue on towards Rome, trying to isolate this breakdown.

At a certain point I called (on a pre-set frequency) our maintenance crew at Fiumicino; they knew me, because I was the flight security representative and had continual meetings with all parts of the company, including maintenance. One of the managers who answered on this frequency was someone I had known for many years; we spoke for a few seconds and he said, 'Look, Francesco, I imagine that the generator won't switch off because there's a switch there that, when it overheats, doesn't work any more.' This means that the breakdown isn't always codified because, unfortunately, being technology-related, something can happen that has never happened before and then the checklist and the protocols don't always account for it.

*So, you're telling me this to argue for the presence of the pilot, while others tell me that, left to itself, the aeroplane would function much better. Like the case of the Germanwings flight, where the co-pilot locked the pilot out and crashed the plane . . .*

Absolutely . . . That's the human factor.

*But he had to deceive the plane, because otherwise it would have taken countermeasures.*

Exactly. Technology, in fact, has tried to remedy these so-called human errors, from the point of view of piloting and technical aspects. Sometimes our senses fail us. When you speak of optical illusions, you understand that we are all subject to being deceived. In the field of aviation, it's obligatory to inform the pilot of the limits of his body. Unfortunately, we have limits and often these limits also influence our decisions: if you perceive a person who's speaking to you in a certain way, and you perceive this communication as an affront, you react. Maybe this person had no intention of offending you, but he just doesn't have a very high level way of communicating.

*No, I thought you meant the physical limit of the human body; for instance, the aeroplane can take measures and perform movements that can be lethal for a human being due to the force of gravity or speed.*

In the civil field it's unlikely, because a civil plane is not predisposed to do these things. We're speaking above all about the mental field, where human error passes through the senses from the physical point of view, because if the brain perceives an optical illusion, it becomes alarmed. For instance, the Air France Airbus 330 that on a flight from Brazil to Paris sank into the ocean: there the human factor played a decisive role. The pilots were not untrained, but there are situations where the senses can deceive us and so we study them. Flight safety goes in this direction.

*They didn't think they were going the speed they were actually going.*

Exactly. As a result, they caused the plane to stall and they crashed.

*We have often spoken about tendencies towards more technology. I have asked whether pilots will lose their jobs due to machines. And you've always said no.*

I know that you handle [technology] in an in-depth manner and above all that the topic of this book is artificial intelligence. If one day we truly manage to achieve artificial intelligence (I personally am a bit sceptical because, for me, intelligence is what man has, composed of a heart and a mind), we'll have to knock down the wall that doesn't allow machines to have a 'heart'. At that point perhaps we can talk about it again. But until we knock down that wall and someone shows me that it's possible, I remain convinced that what man can do, no machine can do.

*For instance?*

Man solves numerous problems per minute. In the example I gave you of the flight from Naples, we resolved the problem mentally, because at a certain point I decided to abandon procedure and shut off the engine before it blew up on me. It was a calculated risk, because man with his brain manages to analyse the situation and manage the risks, handling them and opting for the lesser risk. And a pilot does so constantly. For this reason, there are relatively few accidents when you compare them to the amount that could happen. Because man has intervened beyond the machines . . . Then again, machines are a fundamental help. Without machines, the flight safety system probably wouldn't have reached its current levels. Since the 1990s, the subsequent jump in quality has been linked to studying the human factor.

*You agree that during a flight, the plane carries out thousands of actions you know nothing about; that is, the machine does so many things on its own. But at the same time you feel that you're in command of your vehicle.*

The underlying problem is that currently there are two philosophies for the construction of aeroplanes. One of the main problems with the human factor is precisely this: that very often those building the aeroplanes are human beings. The one designing it is a human being. And he often takes certain things for granted – the interface, for example, which in this case is machine–pilot – which can't always be taken for granted, because the economy pushes heavily for technology to speed up. Technology is surely more economically convenient. The human brain needs knowledge, continual training, serenity, and all this has a price; it costs a lot more. There's a request for everything to be standardized, so that everyone can do everything, even the pilots. But at the same time technology is requesting that the pilot interface perfectly with the new machines. And this, too, is a talent, one different from that which was required before. When I started, they told me that talent began in our backsides – our ability to feel one with the plane. You can see the difference between you, who are a great expert in technology and informatics, and a five-year-old boy: if they hand you an electronic watch, you'll need a minute to understand it, while he needs ten seconds because he's a part of these times. He's grown up, he was born in this era; his brain has been formed in such a way that he's familiar with everything you've had to study. But aviation doesn't always manage to respond to these situations. For this reason, the human factor is a bit of everything – that is, it's the pilot but it's also the one who designs the aeroplane.

*Allow me to provoke you once more. Let's say that your instruments tell you something but your heart, your brain, tells you: 'No, that's not right.' Which do you believe?*

Neither. It's important for it always to be this way. There are two pilots aboard the plane because one has to critique the other. To critique doesn't mean to have conflict with the other, critiquing is something positive, it means comparison. It's a process that allows you to acquire awareness. It's different from judging.

*But let's pretend you're the first pilot, the captain, and on one instrument it says the plane is like this or that, but you look at the wings and say, 'That's not right,' and you touch the instrument and then it resets itself properly.*

It's happened! Just like that! There's a very important factor on board the aeroplane: redundancy. Every critical mechanism has a back-up. For this reason it's unlikely that you'll arrive at extreme situations. Unlikely, but not impossible. For instance, on the Airbus 330 we were talking about before, the situation arose for a whole series of reasons: it was night, the weather was bad, visibility was poor, the captain was not in the cockpit at first and arrived too late to correct the error.

*They stalled?*

Yes. And so they lost lift, which, if we simplify things, is the physical force that keeps a plane in the air, but they were convinced that the instruments were telling the truth, and that they hadn't stalled. I'll tell you another little story. It dates back to when I was first promoted to captain. We had to undertake a course of about ten months and it was highly selective: there were sixty-four of us pilots participating and only nineteen of us were made captain. The last mission I had, before reaching the rank of captain, was carried out on the simulator, and in this simulator I was alongside a second pilot, just someone chosen at random from among all the pilots of the company. During this exam, the instructor presented me with a mechanical failure. After I had identified it, I took the measures required for that situation. The co-pilot wasn't convinced that we had done the right thing. What did I do? I said to the co-pilot: 'Okay, I see you have doubts; I don't have any but according to me your doubts should be analysed because here the survival of the plane is at stake.' And I told him: 'Communicate to the control tower that we're going to stop now at point "X" because we have a technical failure and we have to check it out.'

We spoke with the control tower – in the simulator this was played by the check pilot who was seated behind us. We waited, and I said to the co-pilot: 'Now take the flight manual, because your memory is important, mine too, but since we disagree we need to check . . .' At that time we used the flight manual, not like now where everything is computerized and there are screens and computers on which almost everything you need can be displayed, and the machine even suggests solutions to you directly. Of course, not for every situation. It will take a while for that to be completed, for reasons linked to the interface I told you about: the schools have to change, the selection methods, everything has to change, and it can't change from one day to the next, because since it's a complex system it's highly dangerous to do it all at once. The change is happening very quickly, but it's still not complete.

*Then, we passengers will have zero tolerance with a machine that makes a mistake.*

Yes, of course, absolutely. Anyway, to finish the story, the co-pilot took the manual, and I said to him, 'Open it to page 28 and read the procedure' (because I had learnt almost by heart the contents of these three books that were like the bible for us as pilots). After he read it, he said to me, 'You were right, captain; I was wrong.' 'Okay, we've figured it out,' I said, and added, 'Tell the tower that now we'll continue our flight to our destination.' At that point the simulator stopped, the instructor entered and said to me: 'The mission is over.' I was dumbfounded! I thought I had done something foolish, that I had made a mistake and that he was failing me and I wouldn't be a captain. We got out of the simulator and I put everything in place to show that I had strong nerves and wasn't worried. Actually, my heart was beating about a thousand beats per minute. The instructor looked at me, shook my hand and said, 'My compliments, you're really good!' I could have fainted in that moment, because the tension had completely dissolved. I stammered, 'You didn't let us finish.' And he said, 'Sorry, but I have other duties to attend to and you have to celebrate your nomination as captain right away.'

*So, a month ago I told you that I was speaking with a flight controller at Fiumicino. I think that there, too, they have rigorously selected personnel, because the controllers have our lives in their hands. Yes, the pilot does everything, but for the last five minutes of the flight, the controllers have to align the planes and get them on to the runways. Otherwise there'd be total chaos.*

Not exactly. The aviation system, realizing this kind of problem exists, has invented an electronic device called TCAS.

*TCAS? What does that stand for?*

Traffic Alert and Collision Avoidance System is a device that creates a radar screen around the plane and is able to interface with all the planes that have this system. In commercial aviation all planes have to have it; without it you can't get approved for flight or for the public transport of passengers or goods. Some tourist planes don't have it, and this is the limit that the economy places on technology: for me, either they should all have it, or none of them.

*Yes, that would prevent mid-air collisions. But can it land? You can't say, 'Ah, since I'm a captain, I land when I want.' You have to have a designated slot.*

Absolutely you do, a landing sequence. In normal conditions, when there's visibility, the pilot's visual contact is important. Of course, in landings with low visibility (so-called 'blind landings'), the landing sequence is handled totally differently. Anyway, here, too, the pilot has full responsibility to see the runway and judge whether he's able to land in complete safety. Just think that all this happens when the plane is less than fifteen metres above the runway and going at a speed of about 210 kph. And it only takes the plane about five seconds to travel those fifteen metres. There's no such thing as a totally blind landing in commercial aviation. The pilot is always the one who assumes the responsibility to say, 'This is the right runway and the plane is safe to land.' If he doesn't consider it safe, he's always the one who decides to interrupt the landing manoeuvre and climb. Also, in low-visibility takeoffs, it's the pilot who decides once he's in line on the runway whether there's sufficient visibility to respect the limits set by law; he does this even if the electronic instruments on board indicate a visibility that's different from the one he observes.

*So, the question I asked the flight controller at Fiumicino was whether he thought he could be replaced by a machine and he answered 'never'; instead here you seem to have said yes. How is that?*

No, I said that they are studying a system, which isn't yet operative, that will allow planes to receive authorization for a route that will go from takeoff to landing, through a computer, and everything will be monitored by a PC linked to the radars that will follow the plane through its entire route. There will no longer be a flight controller who tells you that 'you have to follow a particular route', but there will be a group of flight controllers linked to a computer that will handle everything. There will be only one air zone, for instance, one covering all of Europe, coordinating all the air traffic. We can imagine this happening in Europe or the US, but unfortunately there are still areas of the world without radar coverage. Let's take the example of Malaysia Air, Flight 370, which disappeared on a flight from Kuala Lumpur to Beijing in 2014. At a certain point, the flight could no longer be identified by radar or satellite. Even today we still don't know what happened during that flight.

To go back to what I was saying before, I'm convinced that flight controllers in airline transport may vary in number, depending on the type of technology that's available, but until that barrier of artificial intelligence has been overcome, there will always be a need for them. And that goes for pilots, surgeons and workers at nuclear power plants. I have confidence in technology, but I'm not a believer!

For example, in religion, I don't ask questions about the things I can't have the answer to for now, because they're greater than me. I just believe and that's it; this is my faith. With technology, I don't have the same attitude – I need confirmation to believe. These days the mass media are talking about cold fusion. It seems that in America they've done experiments which demonstrate that we're very close to discovering it and in any case it's something that can be achieved by man.[2] Basically, if this happens, it will be a revolution in the world. It will change everything. But, above all, the economic strategies will change, and these are what move the investments in research.

*The last question I always ask is this: do you think that in your professional life you'll eventually be replaced by a machine?*

I personally don't think so. It's hard to give a simple yes or no answer to this question and I'll answer you with an example: when

I was in high school and was seventeen, they taught me that the atom was the smallest thing there was in nature. After forty years, the atom has become, in physics, something as big as a watermelon. How much has changed! So how can I answer a question like that? Inside me, I hope so! But it would mean that technology would have to take huge steps forward. The main thing necessary for that to happen would be for the economy to lose much of the influence it has today in technological progress; in short, it would have to assume a more ethical attitude. At that point – just to lighten the mood – we'll realize it's happened because the insurance policies will be lower when there are more autonomous machines; that is, when the human factor doesn't come into play.

I'd like to ask you to reflect on the human factor by citing an example. I'd like to talk to you about cataract operations done on the eye. Today, it's become an operation that everyone sees as completely routine. No one gives it a thought any more, it's become so commonplace. But have you ever seen how a cataract operation is done? For my job, because I am interested in how the human factor comes into play in complex procedures, I decided to observe some surgical operations. So I'll tell you how the cataract operation is done. The eye surgeon works with a microscope to operate. He makes two little holes, a millimetre in size, into which he inserts a device and instruments to substitute the damaged crystalline lens with an artificial one, all in half an hour. But if he sneezes?

*You go blind.*

You understand? And I asked this of my friends who are eye doctors: what if you have to sneeze? If you have to use the bathroom? And what about brain surgery, which can last up to eighteen hours? They told me that it never happens. This answer reminded me of my experience on board a plane, sitting in my pilot's seat: once I put on my seatbelt I never felt tired, ever. The same thing happens to them. And this is the brain: the machine will never be able to do this, it will never manage to be at our level.

*The opposing argument is that if the machine never has to sleep, it has greater capacities than man. We can furnish it with all the data of all the planes that have ever existed and then the machine can access all this information that a man could never have. For example, the inquest you did at Linate. That could be a file with hundreds of thousands of pages. The machine can handle this easily, and not only this, but hundreds of similar examples.*

I agree, but at Linate the state's attorney, who was a very, very intelligent woman – today she's used as an example in flight security, and not only in flight security but in all the complex systems as a point of reference – she brought to trial all the CEOs of the most important aviation bodies of Italy. The CEOs were those principally responsible for systems that didn't work. I could give you an example. There's a device on board aeroplanes which is like the black box, which is called the ELT (Emergency Locator Transmitter). It activates automatically in the case of an accident and sends out a radio signal on an aeronautical emergency frequency which is the same all over the world. The pilots and flight controllers on duty are obliged to keep it always synchronized. Since I was entrusted with the task of decoding all the radio communications that occurred between planes and flight controllers at the time of the crash – you know, these communications are all recorded by sophisticated devices on the ground – during the inquest the state's attorney asked me why in the control tower they hadn't heard the sound that the ELT had emitted after the crash with the other plane and which I had identified in the recordings. So we decided to make an inspection in the tower and we found that the speaker that was hooked up to that frequency was shut off! This is the human factor, for good and for bad.

*A few pilots have told me that often pilots go ahead even when there's something that could become a problem.*

Of course.

*Many pilots go on, they say: 'Well, let's ignore it'; of course, if they discover that there's no fuel, then they stop, obviously. But even when some instrument fails (for example, the level of fuel isn't what it should be), they take off anyway.*

I spoke to you about redundancy in the aviation field; redundancy helps guarantee flight safety. The pilot knows this. Just to explain myself better, Alitalia at a certain point, to pay less in airport taxes, decided to reduce the capacity of the MD-80s by lowering the maximum limit of weight that could be transported. It's as if you, in your car, which carries four people, were to go to the insurance company and say, 'Instead of four passengers I'm choosing to take a maximum of two in the car', with the objective of saving money. There's nothing risky about this practice at all. But your car is built to carry four. If you encounter someone on the road, and there's already two in the car, and this person needs to be taken to the hospital, your car can do it, right?

*Of course.*

If you do it, what are you doing? Committing a violation or doing a good work?

*Both.*

Well then? That's your answer!

*Once, a pilot told me that he accidentally shut off the engines.*

Here again, we're talking about human error. It's normal. In the presence of an engine failure, if you establish, after having carried out all the required checks, that it won't be possible to start up that engine again, you have to shut it off in order to avoid further risks, because fuel is still circulating inside it. To do so, however, there's a specific procedure that provides for the shutting-off of the fuel valve. In the tension of the moment, it can and has happened that the pilot has made a mistake and has shut off the fuel to the working engine. To avoid errors of this kind, the pilots train at least every six months on the simulator.

I'd like to say one last thing. My experience is that artificial intelligence and technology are taking giant steps and I think we'll arrive at levels that are unimaginable today. History has shown this is possible. I hope it happens and that it happens in such a way that it allows man to live better in the world. Unfortunately, what I think is that we need to take the economy into account. I hope that it becomes much more ethical. It's a bit like trash. We produce a ton of trash, since our life is based on consumption, but all this is also a source of strong economic movements that give rise to considerable individual earnings, and thus power.

*You're insinuating (indirectly) that technology, and even artificial intelligence, could be used against man.*

Unfortunately, the atom bomb demonstrates this, but in my heart, because I'm an optimist, I believe in technology. I've always used it and I hope that it can help man's existence on earth. I hope it does!

*Let's hope so. Thank you, Captain.*

*Dedicated to Andrea.*

## 2.
# *Big Data*
## (*Eric Schmidt and Jared Cohen*)

Upon completing *Futuro ignoto* (*Unknown Future*), my first book of interviews dealing with the impact of the new digital age, I never would have imagined that I would be sitting down with two key people of our time and having a conversation about the digital world. Eric Schmidt is Executive Chairman of Alphabet, Google's parent company, and Jared Cohen is the head of Jigsaw, previously known as Google Ideas. Together they wrote the book *The New Digital Age: Reshaping the Future of People, Nations and Business*, published in 2013. Our conversation, which follows, took place in Rome before their meeting with Pope Francis in January 2016 in the Vatican.

Reading their book was what first convinced me to begin interviewing important people in our society in order to elicit their responses concerning the development of digital technology. The two of them are not only expert analysts of the digital age, but in a very real sense are its architects, too. What I learnt from them was both consoling and, at times, unsettling. I leave it up to the reader to find out the difference.

* * *

JARED: Let's start with 'What has changed since we wrote the book?'

*Okay.*

JARED: I think if you just go to the Introduction and you look at some of the numbers, you can see what has changed. The book was written speculating about what happens when billions of new people come online, and the reality is that this has happened faster than we expected. In this sense, we are seeing some of the hypotheses tested, we are seeing some of them taking place. A lot of things that we talked about in the book have happened faster than we expected. Some shocking things that we wrote about were, for example, WikiLeaks (Snowden hadn't even happened yet when we published the book), and the Arab Spring, which was still very much a sort of open question about what was going to happen. There was no rise of ISIS, which has proven that, as a terrorist group, it is able to occupy both physical and digital territory.

I think we didn't go far enough in the book in talking about some of the ways that technology transforms identity. If you ask Eric and me today about the future of identity, I think we would say that everybody will have multiple personalities because people are basically proliferating identities of themselves, walking around with a virtual entourage. You have your work identity, and you have your family identity; maybe you have your sort of 'Let's behave well' identity. Who you are is an aggregate of all those different personalities. This creates an interesting way to think about identity because we are all increasingly splitting our time between the physical domain and the digital domain. Everything from how healthy we are, to how we think, to who we are and so forth, is an aggregate of all of that.

ERIC: Let me answer the same question by saying that our work was published right before Snowden and right before ISIS, and Snowden and ISIS were both significantly worse than we had anticipated. In Snowden's case, the amount of leaks and the amount of surveillance, the tremendous controversy within the United States over National Security Agency behaviour, was far, far greater than we had expected. Of course, we didn't know the details. America today is still suffering through the consequences of the activities, the leaks, the debates and so forth. That was followed by the rise of this horrific group ISIS, which we argue is the first truly digital terrorist organization. It is inconceivable to any of us (and I'm sure also to our readers) that you would use Twitter, Facebook or YouTube for these videos that are designed to terrify. We have subsequently learnt that these are the recruiting videos for other evil people. We, in our book, did not understand that you could sufficiently terrorize people using the online media to recruit people to your evil organization. There is no question that the internet has aided the creation of this caliphate and evil idea by taking people who are marginalized or unhappy in other countries and giving a way for them to be reached and recruited: something we never anticipated. Both of those – Snowden on the privacy question and ISIS now on the government question – are driving much of the political discussion in the United States and in Europe. You can understand the refugee crisis which is going on today, this humanitarian crisis, as a natural consequence that started with the profound use by ISIS of terror.

JARED: To add a couple of things to what Eric said, let's talk about the States for a minute. To continue down the list of things that happened after we wrote the book . . . One of the things we talked about in the book was that cyber attacks would be so significant as to warrant physical retaliations. Look how the North Korean hack against Sony resulted in increased sanctions on North Korea. To my knowledge, that's the first time in history that we've seen additional sanctions imposed upon a country directly in response to something that happened in the cyber domain. A second example is if you look at Russia's annexation of Crimea, it's an illustration of a point that we speculate about in the book, which is 'We will never see a physical war that's not also accompanied by a cyber war'. So, if you look at, for instance, the distributed denial of service [DDOS] attacks that were happening around the world during that particular time, a disproportionate number of them were targeted at both Ukraine and Russia. You literally see a cyber war between the two countries and a physical war between the two countries, happening in tandem.

*In the book, you mentioned the fact that the future depends on what we do with machines. I think it's in the Introduction, you say, 'Forget sci-fi movies and scaring people; what happens in the future will depend on us.' Would you still reiterate that?*

ERIC: I think we would, and the reason is that you could imagine the technology architecture being changed to mitigate or encourage the kind of behaviours that we have been discussing. For example, you could make it much harder for the government to spy on its citizens. You could also make it much easier for the government to spy on its citizens. You could, for example, require by law that the companies give the government the keys to encryption. There are people who are proposing that. All of these change this calculation. With respect to ISIS, you can imagine artificial intelligence systems that automatically detect hate speech and take it down before anyone sees it. Let me offer an extreme case. Let's say there is a voice, an Islamist preacher who preaches death and destruction to the entire world in a way that is so nihilist that his voice needs to be stopped. You could imagine technology emerging, or being built, that would detect it in its many forms and literally delete it. I'm not encouraging that; what I'm saying is that it is possible technologically. If you believe that ISIS is the source of evil and instability in the world (and many people do), that's an example of something you could build.

*Okay. An interesting comment that my students almost always make to me is 'Professor, why should we have to learn this when we can just google it?' Do you see technology changing the way we teach and the way we educate, the way students interact with digital technology?*

ERIC: In a decade – not now but in a decade – there will probably be a tool that you will have in your classroom that will adapt the teaching to the specific strengths and weaknesses of the students. The way it determines this is that it knows the students, and it knows that some students don't type very well; they may however read very well; they may like to read literature, but don't like poetry. The tool says, 'Okay, Father Larrey, so what are you trying to do?' And you say, 'This is what I am trying to accomplish.' It then says, 'For these students, poetry works, and you need to spend thirty minutes on poetry; and those students hate poetry, but you can get them with a narrative from a movie.' And then you'll say: 'Okay, class, I'm going to talk a little bit, and I'm going to work with the poets over here, and then I'm going to work with the movie people over here.' The computer will help you understand the unique learning paradigm of each student. The reason we know this is possible is that we know that we can pattern-match against people's abilities. Everybody is different: they learn in different ways, so we can train against that.

JARED: The other thing I would add to this, which is not so much a technological answer, is the following. You asked the question, 'What's new in the classroom?' What is new is that students will have more options than at any other time in history. On the one hand, let's take the comment that your students had about Google. As a professor, you think to yourself, 'Does having access to Google remove the influence of the classroom?', and then just send them off on a journey of searching for things online. If we accept that most of the world's population still learns through rote memorization, and if we agree (as I would imagine that we do) that this is a problem, then the question from the student goes more like, 'Why should I sit in the classroom and just learn something that I am being told to memorize when I can go home and find it? Here I finally have an opportunity in my life to engage in critical thinking.'

*Absolutely.*

JARED: It depends on which student you are asking and in what context. This example captures the essence of technology. There is a good story and a more challenging story with any type of technology that we are talking about.

*Many of my colleagues dislike Google, because of that. They think that it leads students to be lazy, that kids don't want to learn any more. I obviously don't share that opinion, but I'm sure you have heard that before.*

JARED: When you say 'your colleagues', I assume you mean other professors?

*Yes, other professors. Sorry, Jared, if I can interrupt, there was a discussion several months ago over whether or not to have Wi-Fi in the classrooms. Many professors said no, because they don't want students using Wi-Fi during class lessons. I said yes, and if students are updating their Facebook profile, then that's the fault of the professor.*

JARED: I think your colleagues who voted against the use of Wi-Fi are missing the long-range benefits of the technology. Look at one of the huge benefits that the internet brings: we have more visibility into what's happening in the world today than in any other time in history. It doesn't mean we are any better at responding to crises, but there is literally not an atrocity that can happen on earth that the world doesn't see. I would argue long term that this creates a demand for us to act. That's how I would answer to your other colleagues.

ERIC: Let's talk about the core question. There is a learning mode in which you sit and your attention is completely focused on a professor's knowledge. You sit there, and I did very well at that (when I was a student). I'm not sure everyone does very well with that. People learn in different ways, and my observation when I go to dinner (which is social and also work) is that people are constantly spouting facts off, which I check using Google. People say something, and I say, 'Let me check that', 'Well, that's okay', or 'Frankly, that's not true'. We were having a discussion about the supporters of Donald Trump at dinner two nights ago. In America, everyone has lots of opinions. Well, I happen to know the facts. I get out Google and I read how the demographics are distributed for Donald Trump. Then everyone says 'Okay', and the conversation goes on. Now, did they learn from that moment? I hope so.

*I think so.*

ERIC: My view of Google is that it is a wonderful source of facts, which can lead humans to think about them: it is a teaching tool, so I disagree with that view that sees Google as a disservice. There are plenty of services which are time-wasting, like Twitter, Facebook . . . these are largely time-wasting in my view. This is my own opinion, because they are essentially social (what are other people doing and so forth), whereas a mechanism for learning and thinking . . . On the flight yesterday, I watched a television documentary on a famous case in America in which the defendant may or may not be guilty, and my conclusion was that he was guilty, so I went to the internet using Google and I read more of the story and I concluded that I was still right, the defendant was still guilty. That's learning. It is supplemented by Google.

*[The interview was concluded here but resumed later with Eric alone.]*

*Can we talk about artificial intelligence? You've been thinking a lot about this lately. Can you speak to us about your thoughts concerning AI? What do you think is AI, are you happy with Google's development in this field, and do you have some concerns about the future?*

ERIC: A couple of things. In the first place, artificial intelligence can be defined as computers doing things that humans appear to be capable of doing. Human kind of activity in a computer. We use AI at Google to do many things. We use it to provide better search results, better advertising, we use it for speech translation, we use it for video and photo recognition, to name just a few examples. At the moment, it is pretty tactical. Think of it like a machine: a car is a transportation system used to transport people. It is an object, a thing. Right now, our use of AI is very tactical: it makes something more capable, making it better and better. We're not at the point where the real questions about artificial intelligence come into play, and won't be for a while.

*And those questions would have to do with, for example, the job market?*

Yes, we're not there yet, we're not having a negative impact on jobs although people are concerned about that. There is no issue about asking whether these things have souls or do they think independently. We're not anywhere near those science-fiction questions.

*Okay, but that could come up down the line?*

You never say never.

*But you're not particularly concerned about that now?*

Not in the short term. Not in the next few years.

*What is your understanding of what we call 'machine learning'?*

Generally, at the moment machine learning and AI are pretty much the same. In machine learning, instead of programming an outcome, you train to obtain an outcome. The simplest example would be this: I want to recognize a zebra. On the one hand, you could write a code that would indicate: 'Look for an animal that has stripes of this kind'; on the other hand, you could show it a lot of pictures of zebras and indicate, 'This is a zebra, this is not a zebra, this other is a zebra, this other is not, nor is this.' The latter is called 'training', the former is called 'programming'. The systems that we are using today are 'training systems'.

*That was a huge leap for digital technology!*

That was a very big gain. That is why these systems are so good at what they do.

*Yes. Are you happy with DeepMind?*

Very happy. As you know, they defeated the Go world champion, Lee Sedol, four games out of five last March in Seoul with AlphaGo. They are making fundamental strides in better and better ways of implementing the algorithms.

*That is an amazing team. I met them in London. Can you address the issue of the 'right to be forgotten', which Google has offered users in the European Community as a response to their obligation in terms of privacy? It appears that Google is being held as 'judge and jury': if you actively suppress links to compromising information, you might be accused of being like Orwell's 'Big Brother'; if you do not, you might be accused of not respecting privacy. There seems to be a dilemma.*

We were forced to do this by the European Court of Justice. The court effectively made us both judge and jury over whether private information should be shown or not shown. As a company we have to implement what the court mandated. We don't have a choice. We have never liked doing this, because it puts us in the position (as you pointed out in your question) of making these decisions, which we think should be made by governments. In other words, Google should not be arbiter of these questions because it is a private company. Nevertheless, we followed the law.

*You may not enjoy doing it, but you seem to be doing it effectively, it seems to be working.*

We are implementing exactly how the court described we had to do it. So, yes, it is effective because we were ordered to do so. We did not have a choice.

*Do you think such a 'right to be forgotten' is going to be extended to other parts of the world, or just remain in Europe?*

That's really a question of speculation, and my guess is that it will not be extended.

*Okay.*

I think each country solves this problem differently.

*This strikes me as important, because people are beginning to realize the significance of being forgotten. This reminds me of something you said in an interview a couple of years ago: you mentioned that you were glad that you grew up in a time when there was no Google, because you could make mistakes back then without everybody finding out.*

Privacy is important, if nothing else because we all make mistakes as human beings. For example, some mistakes I made as a teenager have (thankfully) been forgotten.

*What about teenagers today that make mistakes?*

I can imagine that a lot of those teenagers are going to be unhappy that their mistakes are now on YouTube and on Facebook and on Twitter twenty-five years from now when they are running for public office.

*Do you think that we will learn to be more forgiving of people that make mistakes?*

I hope so, but I'm not sure. At least in the American press cycle, your entire life as a politician is up for inspection, right?

*Oh, yes, certainly.*

I really don't know the answer to your question. It's an important question, but I don't know how people will react.

*Yes, it's speculative. My own view is that we are going to be more tolerant as it becomes more routine to know more about each other, and about our flaws also.*

Yeah. As you know, human beings make mistakes. Any kind of a mistake you make can be used against you by people who are fighting against you using the press. We've seen that in the American political system, and I would assume that this happens in Europe as well.

*Absolutely. It is already happening. I tell my students to be careful about what they post on Snapchat or Instagram. The twenty-year-olds that I have as students seem to be more cautious and aware of the long-term effects. I think that as technology matures, this is what happens.*

Let me give you an example. A sixteen-year-old girl ends up really drunk at a party, and there is a video of her misbehaving, drunk, doing things which no young woman should be doing. That video gets posted. How is that fair to that girl? It's not fair.

*But it happens.*

Happens all the time.

*Are you just going to leave it at that, it's not fair?*

I am certain that it is not fair, but I don't know how to fix it. But I do think it is not fair. I do have a thought on that, though. Especially when it comes to young people, there should really be some tolerance. Young people's judgement is usually not as good as that of adults.

*Do you think sometimes we blame the technology for things which are really caused by human beings?*

Using the example of the sixteen-year-old, someone posted that video.

*But it's not the fault of the technology: it depends on the judgement of people.*

Yes. It is obvious that humans did it, so on both sides. Therefore, you have to hold humans responsible for those judgements.

*Can you talk to us about driverless cars? Google has been a pioneer in this, and now you are teaming up with Chrysler. Do you think it is a safe and mature technology?*

Well, let me ask you a question. How many people die on the highways of the world per year?

*I have no idea. A million?*

The most realistic number I have seen is about 1.3 million people out of eight billion are killed every year. Let's say we could reduce that number by half. That would mean that 500,000 people would now be alive. I believe very strongly that computer-driven cars, self-driving cars, auto-pilots, all of that technology is very useful for automobiles. We and other companies are trying to get that to happen within the next year or two. They are not going to be perfect, but there will certainly be fewer deaths.

*Do you think that people are going to be okay with that, 'they're not perfect but there are fewer deaths'?*

Again, we have tolerated 1.3 million people dying in car accidents, right? Just here in the United States (I don't know what the number is in Italy), there are going to be 33,000 people that die on the roads this year.

*That's a lot of people.*

That is a lot of people. How many US soldiers died in combat last year?

*Far fewer.*

Fifty, one hundred, some number like that. Self-driving cars are a very big deal.

*You mentioned 'a year or two', so you think that the technology is mature now?*

The technology works most of the time. It's not completely perfect, but it is very close.

*You're saying that we can save a lot of lives through this technology, which seems obvious. Are some people afraid of driverless cars?*

I'm sure they will be, but when they understand that they are safer, they will get over their fears. For example, some people go to the bank, and they want to talk to a human and not use the ATM. That's fine. There will always be cars which you can drive as well. But if you want to get the number of deaths down, you will want to use self-driving cars.

*Are you going to give these cars a kind of ethical system, or a set of rules?*

They have a set of rules. In general, their job is to protect all the humans. That means the humans in front of them, humans in the car, etc. We have heard of these scenarios in which the car is faced with a pedestrian on one side, a dog on the other, a child in the middle, and how to make the right choices. But those scenarios don't really happen.

*Okay [laughs]. Because those are really interesting ethical quandaries.*

When we get to the point where that is the hardest problem, we will deal with it. But right now, we're just going to stop the car quickly.

*Exactly. The more driverless cars are out there, the safer we will be. The problem is then going to be real drivers who are unsafe.*

Yeah.

*Where is Google headed in terms of healthcare? Can we expect some interesting breakthroughs in the near future?*

We are now working not on drugs, but on devices. We are very interested in medical devices that can help monitor your health; we have built this contact lens that can measure glucose levels and a person's blood state. We have partnerships of that nature in the group called *Verily*. It's a long process. We say, 'Let's invent helpful tools but let others commercialize them.' Our side is the Research and Development shop and then we use computer technology to make medicine more reliable. We have projects, for example, in cancer analysis, trying to use large data and DNA databases to help cancer patients. We have some doctors and a lot of programmers, as opposed to a lot of doctors and a few programmers, which is what everybody else has. Our niche will be that.

*You conceive Google as more like an 'incubator'?*

For healthcare, yes, because we don't have the ability to do large trials and we are not a big healthcare company.

*Yeah, that requires a lot of money and time.*

Maybe in the future but not in the short term.

*People tend to look to technology as a type of 'saviour', don't they? As the technology gets better, we live longer.*

There is no question. When you are eighty years old and you have some type of cancer diagnosis, you really do want the drug which will stop the cancer. The way those drugs will be invented will be largely on the basis of DNA analysis, new algorithms, new studies and a lot of risks in the medical profession. In your case in thirty years from now, you will be happy that they did all that work.

*Absolutely.*

My grandfather had a heart attack at sixty-five, and today in America heart attacks are relatively rare in the young age group. The mechanics of heart disease and by-passes and so forth are well understood. All of that is due to the research that has been going on in the last twenty or thirty years.

*We are going to be seeing some amazing things come out. You talked about the device for measuring glucose for diabetes patients: I have friends who are diabetic and they would love to be able to use something like that.*

Diabetes is a really unpleasant disease to have, so can we help fix that? I hope so.

*What about 'big data'? Is it a friend or a foe? Can the information that companies like Google and Facebook have about us be potentially disruptive?*

Big data is a reality. The computers which we use every day naturally collect that data. That raises certain questions like who is using the data, and what are they using it for? I argue that Google uses big data to provide valuable services, and if we were to violate your privacy, you would stop using us. More importantly, we would be sued by privacy activists. There is always this question of large corporations and the data that they assemble. Corporations have many reasons to keep that secret. I am actually more worried about the governments' big data, because governmental systems tend to be poorly architected and easily broken into. In America, we have the Office of Personnel Management which has a lot of people's files; a bunch of IRS databases were leaked by hackers (who are criminals). I am sure the incentives for Google, Facebook, Yahoo, etc. are to never allow that to happen. They are not going to be perfect, but they work well.

*Yesterday I was reminded that it's against the law to steal someone's Social Security Number, or to use it without the owner knowing. But in reality my cell phone number is more important than my SSN. Do you think that is accurate?*

That's very interesting because nowadays your cell phone number is becoming more important in terms of your identity. It is possible that your cell phone number will become your primary means of identification. In the American system, Social Security Numbers are not a legal ID in the sense that they were not issued as a national ID. They are used in the financial industry for banking and also for tax compliance. But I don't use my Social Security Number every day, except for financial transactions. But I use my mobile phone all the time, so I agree with your point about cell phones being more crucial.

*Do you find that there is collaboration among the industries based on data sharing? Let me ask you specifically about insurance companies. They have access to a lot of our information, and they make insurance policies on the basis of analysing big data which humans can no longer process (because of the amount of information). Do you think that is a positive trend?*

I think it's probably okay. Again, we are careful about data sharing for precisely this reason: we don't want our data to be leaked by others. There are some natural limits to data sharing. Most people want the government to share the data, and they do not want companies to map the shared data. I'm sure there will be more and more restrictions on all that.

*I think that's the tendency. You are right. Two personal questions to conclude. Why did you want to talk with Pope Francis? What can you tell me about that meeting?*

Well, you were there, so you can recall what we talked about. I travel the world talking to leaders about the internet, and I had no idea what the Pope's view of the internet was (so I wanted to ask him). He is obviously a very thoughtful man, and my impression was that he cares a lot about how it is affecting the way people interact with each other. He commented that he is worried that people are no longer having dinner-table conversations because they are all on their smartphones or tablets.

*Right.*

He's concerned that this technology be available to everyone, not just the elite. He listened to my view about the spread of technology. If I can help His Holiness and you guys in other ways, let me know.

*We are working with the YouTube people and trying to see how we can best use that avenue to transmit the Pope's message. It takes time, but we will make it happen.*

I am a strong proponent of getting the Pope a two-billion-viewers TV station. Everyone who is Catholic can see the Pope and humanize the Pope and his message. It's just a new technology to get the Pope's message out, a message that's been true for hundreds of years.

*The Holy See is very grateful to you, because the Pope has an amazing image and persona in the media, yet very few clicks. So, we need to do something.*

There is obviously something wrong, something we can be doing better.

*Here is the last question. I think you've been asked this before. What motivates you today? By almost any standard, you have achieved so much. What, in your professional life, still makes it meaningful to go to the office every morning?*

My own view of life is that you have to make a difference in some way. That is why God put you on earth, and you have to maximize whatever skills you have and you need to have a good time doing it. A combination of impact and enjoyment is the best life, and if you look at health and happiness and income levels, it is all correlated with: 'Are you doing something you find meaningful, and are you happy doing it?' I am in the bizarre situation where I can decide how I spend my time, and so I care a great deal about spreading the message of the internet and the empowerment of individuals. There are always issues. To me, at the end of my professional career, to be able to have had the kind of impact that I have is very satisfying. Remember, I started off as a programmer.

*You were a computer engineer?*

Yeah. It's very fulfilling. As an example, yesterday (as you know) we met with President-Elect Donald Trump here in New York, along with other leaders of digital technology. Well, I asked myself, 'What am I doing in this room?' I'm just a boy from Virginia [laughs].

I try and remind myself every day that I am very lucky: opportunity came my way and I took advantage of it.

*I guess it was a combination of being the right person at the right time. It really comes down to motivation. You are motivated to continue doing what you are doing.*

Yeah. Wouldn't you?

*I consider life a vocation. Although you didn't use the word 'vocation', you implied a similar concept. I am also fortunate enough to love what I do. I don't do it as a chore, because I have given my life to being a priest.*

If you think about it, you get up in the morning, and you do what you want to do. In your case, you are serving your religion, you're serving your students, you are an intellectual, you are writing a book. Sounds like a pretty good deal to me [laughs]. Right?

*Well, I think you're a better writer than I am. You're absolutely right. I do know many people who grudgingly go to work every day, and they don't enjoy it but they have to do it to make an income.*

Is there some way that we can make a difference? I don't know any other way to say it. Everybody serves the world and their community in a different way. This is how I do it. There is a lot of correlation that when you stop, you die.

*Absolutely.*

It seems to me that we just say yes.

*That's a great way to end this interview: a significant message that people are going to take away from reading the book.*

People will read your book. Let me know if I can help. I want to stay with you guys, I want to stay connected.

*We will. Thanks so much. Take care now.*

## 3.
# *The Future of Design*
*(Don Norman)*

Struck by an article which presented Professor Don Norman as the Silicon Valley guru who has influenced digital technology more than any other, I was grateful to have been able to meet him after a speech he gave in Modena, Italy, on the role of design in automation. Don is Director of the newly established (2014) Design Lab at the University of California, San Diego. He is also co-founder of the Nielsen Norman group and an honorary professor at Tongji University (Shanghai) in their College of Design and Innovation. Somehow, he also finds the time to serve on the boards and as advisor to many companies and organizations.

Given his role at the forefront of modern design thinking, I was keen to discuss with him the concept of design in terms of new technologies. After our conversation, I became more optimistic about what lies ahead for us, particularly due to his conviction that it will be us humans who will shape and guide the process, ensuring that technology continues to assist rather than hinder us.

* * *

> *Thank you so much for accepting this initiative of mine and conversing with me. I would like to begin by just asking if you could give us a brief biography of yourself.*

I started off with electrical engineering. My first degree was earned many decades ago at MIT doing electrical engineering. Then, I went on to get a Master's Degree in Electrical Engineering at University of Pennsylvania. But I was really interested in computers and computation and human thought – and the Electrical Engineering

Department was not geared for that. At that time, the Psychology Department changed direction and the new head of the Psychology Department was actually a physicist and so I went there, where they were developing a new field called mathematical psychology.

The head of the department said, 'You don't know anything about psychology. Wonderful. Come join us.'

*That's funny.*

In any event I got the degree. I decided that doing psychology was the same as doing engineering except instead of trying to build intelligent systems I was trying to understand them. Now, in those days psychology was ruled by behaviourism. It had no interest in studying the mind or how the brain and mind worked. So, I was really a newcomer. And that made me both an outcast, but also at the forefront of the newly developing field. When I graduated from Penn, my first job was at Harvard University and I do remember I really started to work on trying to understand the mechanisms of the mind. I was denounced by B. F. Skinner at a faculty meeting: he denounced me and my field. It was fun [laughs].

*That's amazing.*

But after five years at Harvard, I joined the newly started university, University of California, San Diego, and I got there before any students had ever graduated. The chair of the department had set up something called 'The Center for Human Information Processing' and that's what I did, human information processing. This was like today's current digital era affecting psychology then. The notion is not that the mind is a computer, it is certainly not. But rather we should try and understand the mechanisms that are inside better to understand people. Thus, I was at UC San Diego for twenty-seven years before I retired, which happened in 1993. And I helped develop the field of information processing which then became the field of cognitive psychology.

I then decided that psychology was too restrictive, that we needed to combine neuro-science, and linguistics, and philosophy, and sociology, and anthropology, and computer science, and so we started a new department called Cognitive Science. I started the first department of cognitive science in the world, I think.

*So you were really a pioneer in that field?*

Yes, I was, but there were many of us, of course. You can't start a department without demonstrating that this is a real field. I've just felt mainly I was lucky, because of my engineering background. And it was the right timing. Interestingly, I retired in 1993 and went to Apple and became the Vice-President of Advanced Technology.

*Okay.*

That's where I really discovered the importance of design. Actually, even at UC San Diego, I said, 'You know, given my engineering background combined with psychology and cognitive science, design is perfect because design is all about making modern technology appropriate for people.' You need to know both.

*That's interesting. Can you just briefly describe what you mean by design?*

Many people think that design is about making things look good and pretty. No. That's not what modern design is. Modern design is a way of thinking. It's a way of thinking and the field I call human-centric design is thinking based around people. We first try to understand what the real needs are of the people for whom we are designing. Second, we try to make sure we're solving the correct problem, which is almost never the one we're told about [laughs]. Through observations, we try to figure out what is really going on. Third, we believe in continuing, repeated testing where we have an idea, and we quickly make it. By making it, I mean we might do it just by a drawing, or use a cardboard box or foam, and we see if that's what the person needs, and we change. We are continually iterating and changing. We make sure that what we are doing really fits certain needs and, moreover, is understandable by people – understandable and enjoyable.

I wrote a book before I got to Apple, which was originally called *Psychology of Everyday Things*. It later got renamed *Design of Everyday Things*. In Italian it's called *La caffettiera del masochista* ['The coffee pot for masochists']. It's actually very popular in Italy. This book was not well received at first, but it has gone on to sell 600,000 or 700,000 copies. I just released a new edition two years ago, and it's doing well. It's really saying, 'Let's try to make sure that things we build fit people and, moreover, that you can understand what they are'.

Of course, one reason I went to Apple is that was Apple's philosophy as well. When I see four-year-old children with an iPad, who move around completely at ease with it, I think that's a testimony to Apple's design. They did it.

After Apple, I did a bunch of other things. I started a design programme at Northwestern University. I retired from there. Then, I was living happily retired twice, enjoying myself on many company boards, very active, giving talks around the world. But the head of the University of California, San Diego, asked me to come back and start a new design lab. Which I did.

*Okay.*

He gave me the instructions that he didn't care what I did but it must be important and it must be exciting.

*And you had already been retired at that point?*

Twice retired.

*Twice. Okay.*

Now, let me tell you what I'm doing because it's highly relevant to your quest. I want this to be a major design lab, and therefore it has to be different from what the other ones do. There are lots of good design schools, but I want to emphasize the thinking and where the needs are, and the needs I think are around two issues: one of them is the role of automation. I am very annoyed by the way automation is being developed. Primarily, engineers automate whatever they can and leave the rest for people, and when things don't work, the people are blamed for making errors and I say, 'No'.

What has happened is that we people, we're very good at being creative, we can adapt to new circumstances, we are very flexible, and that's wonderful. People say that we're always distracted. That's the typical way that the engineers think: you put a negative term on one of our virtues and our virtue is that we're exploring the environment. When new things happen, we attend to it, and we're very good at learning about all the different things that are happening and saying, 'Oh, yeah, that's interesting. Oh, that one is interesting. Oh, you know, there's a great combination of those two.' The engineers like to say that we're being distracted. No. We're doing what we're good at. The problem is they build machines that do not allow us to do that, that require us to be very precise, to be very repetitive, to make correct decisions very quickly, and never take our attention away from the machine. Those are all things essential to the machine, and all things we are bad at.

*That's absolutely true.*

That's one of the things I want to address in this new department. I also want to look at complex social technical systems like healthcare, medicine, education, transportation systems, the environment. These are complex issues. They probably do not have solutions, that is, not a single solution. But they are being addressed by people with MBA degrees and engineering degrees, thus making things more efficient and increasing productivity, reducing waste, yet without any consideration for the impact on people. My goal is to get back to having an impact on people, that we must take into account people's virtues; what I want is human-technology collaboration, not this mindless automation.

*Excellent point. I wonder how much you think market forces enter into those kinds of engineering decisions.*

Market forces are there, but the market force is now minor. I believe that a collaborative approach could be extremely successful, that it's not a market issue. I'm really not a fan of traditional analysis and the way that MBAs are trained today. The way that market forces are analysed is all wrong, it's all backwards. The notion that a company owes its loyalty to its stockholders is short-sighted. If you actually look at where that comes from, you will see that it was just some professor's opinion that got captured by everybody. We don't owe our allegiance to our stockholders, we owe our allegiance to the community in which we're embedded, to our employees and to society. We can do all those things and still be profitable, that's not the issue, but it's where you put your emphasis.

*Exactly. Could you give me a brief synopsis of some of the ideas that you spoke about when you were here in Italy?*

I gave talks in Milan about this, about the role of automation in automobiles, but again from the human point of view: about how we interact with these devices. If I want to cross the street, and there's an automatic car, how do I know it's seen me? People aren't thinking about this. How do they interact with each other in a way that gives us trust? I actually think that automated cars are going to be a good thing, for the kind of reasons that I've stated, because driving is one of these things we're not that good at.

*Our track record isn't that good. No.*

Everybody thinks that they're a great driver, but if you look at the accident rate, the United States has 33,000 people killed a year, a million injured. If you look at Italy, it's even worse on a per capita basis. This is because when you first learn how to drive, it seems impossible. But after you've been driving for a year or two, it seems easy, and that's what's really bad because people don't pay attention any more and their mind wanders. As I said, naturally, because that's how the mind is designed. That's how the mind works. It wanders.

Second, people drop things on the floor. They bend over and pick them up while driving. We've seen weird behaviours. The trouble with driving is that it is mostly easy, but when something goes wrong you must respond in a second or two. In one second at 100 kilometres an hour, you've gone 30 metres.

*That's scary. I spoke with the director of Google for Europe, Middle East and Africa. And he said that insurance costs will go down if you have a self-driving car.*

I just wrote an article called 'Automatic cars or distracted drivers: We need automation sooner, not later'.[1] In that article, I said, 'Automation is imperfect today, it's got a long way to go before it really understands how to work in crowded cities with a lot of people, but it will decrease the number of vehicle-related deaths.' The accident and death rate in automobiles is very high. Second, the distractions caused by other equipment – radios, tweets, email, all this is increasing in the automobile, and therefore I said, 'I think that we should go to automatic driving as quickly as possible', because even imperfect automation will be superior to today's imperfect driving. In addition, automation will continually get better, whereas distractions will continually get worse.

*You're absolutely right.*

And will insurance rates go down? Yes, certainly. If the accident rates go down, insurance rates will go down, maybe. The problem is that in the United States, we have a litigious society, so when there is an accident with an automatic vehicle all the lawyers will rush to the scene and get obscene amounts of money. Even though those same accidents are occurring all the time with normal drivers and no one pays much attention.

*That's also true. You know, we may work that out as a society, the more we become accustomed to the self-driving cars.*

Well, what I've always argued is that we absolutely will work it out but it may take decades. And the transition is the difficult part.

*Yes. Would you like to share some of your thoughts on artificial intelligence? That seems to be quite a polemical issue these days.*

Well, I think it's a polemical issue for two reasons. First, it is badly misunderstood, and second, there have been a few people profiting by it, by making it polemical. I consider myself partially in the field of artificial intelligence, for many years I've published in the journals and so on. One of my students was Geoff Hinton.

*He was in the news just last month; he gave a talk in London.*

Well, he was one of my post-doctorate students many years ago, and actually the connectionist view (the neural network approach that he is championing) was developed in my laboratory.

*That's amazing.*

Geoff Hinton was working with my collaborator, David Rumelhart, when Geoff went on to pioneer what's now called deep learning, which has been the basis of many of the advancements in speech understanding and navigation, and basically machine learning. AI is not as powerful as people think, and I'd like to enlist it on my side. I'd like to enlist it so that we have intelligent assistants: competent assistants that help us do things. But there's all these people going around and writing books, saying AI machines will get smarter and smarter and then they'll learn how to build themselves, and once they learn how to build themselves, then it's the exponential increase, the singularity. Machines will outsmart humans and then it's the end of human life as we know it. We'll be pets kept by machines.

*You said a lot of very profound things in that short amount of time. There is, I think, an argument to be made, is there not?*

Yes, there is and actually I'm collecting a lot of the papers; I've started writing a few papers myself on this. And I'm about to teach a graduate seminar on this topic. Graduate seminars are basically how professors learn, that's why I'm doing it: I want to examine these topics in great depth and having students who are smart, who can read and argue and debate and bring in new points, will help me advance my thinking.

*Have you come across Nick Bostrom's book on superintelligence?*

Here we go [chuckles]. Yes.

*Okay. He's probably one of the people you were referring to before.*

He's one of the people I was referring to, because his arguments are all very logical, all very sensible, but they start off with some funny assumptions. I find this a lot, that people make logical-sounding but wrong assumptions, then they develop a long argument, and then people usually examine the argument in detail, and they can't find the flaws, so they accept the conclusion. They seldom think about whether the assumptions are right.

This is what happens in economics. The economists do brilliant mathematics and do all sorts of wonderful proofs and stuff, but they make some basic assumptions, like, 'Oh, if we have two products and they're exactly equal and one is more expensive than the other, people will always choose the cheaper one.' Well, that turns out to be false. There's an axiom to them, they never examine it. Two of my friends (one unfortunately has died) – Amos Tversky and Danny Kahneman – have spent a lot of time demonstrating that the basic assumptions of economists are wrong. I'm delighted that their work has now received a Nobel Prize, and has started to be accepted. Unfortunately, Amos couldn't get the prize because he died, but when Danny received it, his first statement was, 'This is a prize for the two of us.' It's called behavioural economics, and they're finally starting to get more psychologically sound axioms. But Bostrom, I think, is not on the right track – his basic fundamental assumptions are wrong, and so there are many other possible courses of action.

> *Okay. I know you have to go soon, so do you think that what we consider human nature can be altered through digital technology – through this revolution that's taking place?*

That depends a bit on what you mean by human nature. First of all, if you mean the biological structure, I am concerned about that. Building the machines is not going to change our biological structure, but as you know, we are starting to implant artificial organs, we're starting to talk about designing the DNA structure and inserting new types of DNA and so on – that's something very dangerous. Let me just talk about some recent neurological technology. It will affect how people think and what they do, but let me tell you that everything does.

I'm always amazed when people say, 'Oh, we studied taxi drivers in England and we discovered that the part of the brain which does spatial organization is bigger than in non-taxi drivers and it got bigger as they learnt the streets.' And I say, 'Well, of course.' When you learn something and you behave differently than before you learnt it, where do you think the learning takes place? It's in the brain, so the fact that you notice a change in the brain becomes trivial. It would be surprising if you did not notice any change in the brain.

One of the biggest changes in the history of humankind, in terms of technological change, was the invention of writing. In terms of how people think and how they learn, writing was the biggest invention. Then the next big step was basically inexpensive printing, which in Europe was the Gutenberg press. Before that, writing was restricted to the elite, and books moreover had to be hand-copied and handwritten. Once the press got started, then all sorts of things came out. It's interesting to look at what was first published. A lot of them were handbooks: handbooks for agriculture, handbooks for technology, that is, tools for agriculture, for keeping time, for wells, for roads, for carts and, unfortunately, of course, a lot of the tools were for the military.

*Unfortunately. That hasn't changed much over the years.*

I believe this did cause human nature to change in many ways. For the first time knowledge could be much more cumulative, in a world where people could communicate over great distances. It was like I'd write my ideas and people a couple of hundred miles away would receive them. If you look at the travel that went on in early Europe, I am just amazed at how the scholars in Italy were corresponding and working with scholars in France and Germany. A tremendous amount of interplay was achieved. This is what's happening today, in an order of magnitude improvement, that is. Look at us: you and I are talking over a digital line. This has enabled us to maintain trends across the world and to work together. I work with collaborators in China, in Great Britain, in Italy, in Germany, in South America, across the United States, and, sure, this changes the way I think and the way I interact. I don't know if that's what you mean by human nature, but I think it's an important change. I think it's a positive change.

*Yes, that was another question I had; especially the way we're communicating. Some people say we're too dependent on our cell phones, we're constantly attached to them and we're losing the ability to just relate to other people in person.*

I don't see that at all. There are two issues. First of all, cell phones are relatively new. That is, you measure it in years or a decade or two. In fact, it's not even two decades, it's one. So you have to really wait a much longer time to see what the impact is. Second, things like Facebook and tweets: I've discovered they're very powerful for keeping in touch with people and I think that makes my relationship with people far better. Yes, it's very important for me when I'm talking to someone like you that I'm attending to you only and not being distracted by this other stuff. It's true that many people are too distracted, they can't continue a conversation, but I believe that's a temporary phenomenon.

*What do you see among your students, because they are a good cross-section of humanity to consider?*

I find my students are brighter than they've ever been. I have been amazed at the quality of the students and I simply tell them they can't use their devices in the classroom. I say, 'It's my job to keep you interested', but I also know that if it's open they can't resist it.

*They're updating their Facebook profiles.*

Well, it's not that. It's reading other things not related to the class. I have discovered that I do that myself. I go to a lecture that I've been looking forward to, and it's interesting, and if I have my laptop out to take notes, I might get an email and say, 'Oh, a new email that's just arrived,' and I start reading, and then I catch myself and say, 'Wait a minute. I'm here to hear the lecture.' It's just hard not to.

But the students all respect that. At least my students do, and I have a wide range from beginning freshmen to post-doctoral fellows who are very advanced, so I'm very optimistic.

*It's good to hear you say that. I have interviewed people who have taken a much different approach, but it really depends on what one's experience has been and yours has been overwhelmingly positive, I think.*

It's also interesting when you make a statement in class, and then a minute later one of the students pipes up and says, 'No, that's wrong.' And then I say, 'What?' They respond, 'I just googled it. Look.' I think it's kind of neat actually.

*Isn't that scary, though, for an educator?*

The answer is no, it shouldn't be. So many educators think that, 'I am the font of all knowledge, and I present to the class, and I know everything and if someone dares contradict me, it's frustrating, or horrible, or terrifying.' Yet I say, 'No, I'm learning along with the students, and we're doing this together.' If the student is going to disagree with me, I only care that it be an intelligent disagreement, because in that disagreement the class is going to learn a lot and I will learn a lot. If I prove the student wrong, well, the student has brought up a point and maybe I wasn't clear, so I will know how to be clearer next time. If the student demonstrates that I'm wrong, then I am wrong, or maybe incomplete, well, I've learnt something, that's good.

*It takes a lot of humility to get to that point.*

It is humility but we all should be honest and recognize that nobody knows everything. Again, if you think of yourself as a mentor and as learning with the students, it doesn't even require humility. It's just a different mindset. We're so used to the fact that the teacher stands in front of the class and lectures, but actually lecturing is the worst way to teach. It's the easiest for the teachers and it's the worst way to learn.

So we in our laboratory refuse to talk about teaching, we only talk about learning.

*I think we're in for some interesting changes in terms of how we teach – okay, how we learn. Look at the interaction, for example. I mean, I'm already seeing it with my students also. You've noticed it. What do you think is the long-term trend, if you can address that?*

The long-term trend, I think, will be very interesting. If we continue the way we're going, there will be an ever greater gap between the elite (in terms of jobs and salaries), and the rest, whose salaries will go down, and the jobs become either more difficult to find or less interesting. I'm very scared about that. Furthermore, I am worried about the way that automation is taking place, and the way the current economic models of capitalism are encouraging this, as if responding to a powerful elite in the world. Certainly, Italy has seen a lot of that but so has the United States. I've even seen this happen in China, where the elite really control everything, and I think that's a dangerous direction.

But there's another more encouraging direction and more and more people are speaking of this and trying to make a more humane technology, so that it helps people: makes them more creative, makes the jobs more interesting. We can bring back craftsmanship. The fact that we have cheap 3D printers, and laser cutters, and digital tools that make it easier to design things and share them across the world might bring back the rise of individual craftspeople at all levels of crafts: from making physical objects to helping with conceptual abstract ideas. Crowdsourcing is an excellent way of bringing people together. I think that if we start designing our technology with a notion not of replacing people but of enhancing them, there will be a very positive outcome. These are two extremes, and usually when you have two extremes, what really happens in the end is something along the middle. I want to spend the rest of my time pushing for the harmonious collaborative interaction with technology.

*Your worries are very, very important and realistic. I also hope that the wiser people prevail and opt for exactly what you're suggesting now.*

I'm very reassured that there's an increasing number of people who are starting to make similar statements. Many of them don't put the human collaboration there because that's not their background, but everything they do is consistent with what I'm hoping for. If we can get a larger and more influential group of people believing this, then I think we will meet our goal.

*That's one of the reasons I'm writing this book. It's been a fascinating conversation. I really wish we had more time, but thank you so much for sharing these insights with me.*

Okay, good talking to you.

## 4.
# *Military*
## *(Elliot Rosner)*

One of the sectors of society that has been most profoundly impacted by the digital revolution is that of the military. It is safe to assume that advancements in technology in the military are years ahead of civilian developments. We all remember that the first GPS systems were created for military use, and the armed forces continue to lead the development of many other technological systems, such as drones, automated weapons, communication encryption software, lethal satellites, spying devices and so on.

One use of digital technologies in the military sector which has proven very helpful is that dealing with simulated combat systems. Retired Colonel Elliot Rosner agreed to be interviewed about the technology that his company, Cubic, sells to many different national militaries. I was intrigued to learn about how many digital technology systems are being used by the military *not* to take lives, but to save them.

* * *

*Can you give us a summary of what you have been working on over the past years?*

I have been in the military for thirty-five years, and during that time I did many tours around the world. I was in Ethiopia and then I attended the United States Military Academy at West Point from 1972 until 1976.

*You're a West Pointer?*

Yes, I graduated and went to a series of schools and then I served in Berlin. I've served in the Netherlands; I've served at Fort Knox, Kentucky; Fort Bragg, North Carolina; Vicenza, Italy; then at the Pentagon. I served in Fort Leavenworth, Fort Lewis and at Guantánamo Bay . . .

*You've just been everywhere!*

And then Honduras, and my final assignment after the Pentagon again was in Rome, Italy, where I served and retired in 2006 as the Army Attaché at the American Embassy in Rome. I was married for twenty-seven years to a former officer and I have one child, a daughter, and she has given me two beautiful grandkids. I live in Rome with my corgi.

*And why Rome?*

Well, when I retired from the military in 2006 one of my previous bosses worked for a company in the United States called CACI. They were in the military contracting services, so he brought me to the United States and I was doing international business development for him. He then left the company and went to work for Cubic, and when Cubic won the contract in Italy, he said to the CEO and the president of Cubic Defense at the time that they needed to have somebody in this country and he believed I was the man. I interviewed and got the job and I've been here now for four and a half years.

*You mentioned Cubic. What is that?*

Cubic Corporation is headquartered in San Diego. We design, integrate and operate systems products and services that increase situational awareness and understanding for our customers in the transportation and defence industries. Now, what that means is that we solve complex problems with innovative technologies, systems and solutions. We collect and analyse large-scale data that is usable to our customers. We make it our mission to increase efficiency and reduce costs and improve the user experience and we accelerate the implementation of next-generation solutions. With a presence in nearly sixty countries and over six decades as an industry leader, Cubic has established a global reputation as an innovative and trusted partner for our government and commercial customers worldwide.

> *Okay. What we're trying to do is ask different people in different sectors of society how the digital revolution has affected their lives. Can you give me some examples, concrete examples, where you see the use of technology in your line of work?*

Cubic is comprised of two different companies. One is Cubic Transportation and the other is Cubic Global Defense. I work for Cubic Global Defense. I can also talk about Cubic Transportation, but at Cubic Global Defense what we do is provide realistic mission-centre training services, innovative training solutions, C4ISR, which is command, control and computers intelligence surveillance and reconnaissance systems, intelligence and cyber solutions for the United States and allied forces. What exactly does that mean? We provide training systems and services. A concrete example that you might be asking for concerns major systems for which we are the world leaders. The first is a system called MILES, which stands for Multiple Integrated Laser Engagement System. This is basically a laser 'paintball', where soldiers are equipped with a series of sensors, and a harness with sensors on it that they wear on their body. Their weapon has an actual laser system attached to it so that in a simulated engagement, you would shoot an individual, and if you aimed and your weapon was properly zeroed, and you hit the individual with the laser, then his device would go off providing an audible signal so that he knew that he had been hit. He would then have a modular display that would tell him whether or not he was killed or if he was wounded; whether or not he could still move or talk; whether or not his weapon still worked, and so forth.

Therefore, you can imagine an environment where you are training larger units: as they are deploying and conducting their exercise, you would have both 'blue for' (that would be friendly units), and you would have 'op for' (that would be opposing forces). As they engaged in a battle, you would have casualties on that battlefield, and you would have to deal with those casualties accordingly. This system increases the fidelity and the data collection that we talked about, and you can instrument this. By instrumenting you would attach a radio, and that radio would then talk to a communications mast and relay that information from the battlefield or your training area back to an exercise control room, where the terrain and the actions of each individual soldier could be monitored in 2D (which is a way to view that individual); or in 3D where you add cameras and video the exercise. Here, you would see each movement and each exchange of their weapons or discharge of their weapons and could record the actual radio transmission, so that you could see the movements, the indirect or direct firing and engagements as well as the communication interchanges that go on between the individual soldiers and lower to higher units providing orders.

What is the importance of that? What is the added value of that? Having commanded at platoon, company, battalion and brigade levels, I can tell you that it is difficult to see how everybody in your unit performs. What this system provides is that all of your units can be tracked, and it gives you the ability for an after-action review to sit down and see what everybody is doing. If your soldiers are not following your tactics, techniques and procedures, we can monitor that and then we can educate the soldiers, retrain them, and increase the readiness of those units. The added benefit of this is that we can create any scenario that we want and allow our soldiers and our staffs to go through those scenarios prior to being deployed in a real combat situation. Now, you don't use real ammunition, you use blank ammunition, but it allows you the ability to do exercises over and over again and to learn at an exponential level the abilities of your staffs and your individual soldiers. They come to understand that if they had failed or deployed poorly, in a real combat situation we would be having memorial services for them. It increases a soldier's situational awareness, and increases their individual participation and interest in learning how to do things the right way.

*If I can quickly interrupt: how many soldiers can you deploy with this simulated system?*

The system can easily have up to 2,000 or 3,000 people at one time. We are talking about fairly large tactical forces. The Joint Readiness Training Center at Fort Polk in Louisiana is an example of this.[1]

*That is a lot of people and no one gets hurt really because the weapons are simply lasers, right?*

They are laser weapons: your system goes off and then obviously we force the casualty evacuation system to work, because if somebody's wounded and the medics do not do their proper procedures to provide triage first aid, then your health diminishes until such time, after two hours, if you have not been taken care of, then the system will kill you. That becomes another item for discussion in the after-action review as to why finding casualties and evacuations had not been taken care of. The umpires (they are called observer coach trainers and are usually military personnel that observe the actions of the force that is going through the training, and they talk to the exercise control people where the analysts are and who are watching the battle as it ensues), the two of them talk so that the proper things are captured on the battlefield in order to provide feedback of the event and state the lessons learnt. Now, they have what is called an umpire control gun, and it is actually called the 'God gun'.

*And what does that do?*

With that device, once a soldier is wounded or killed in the simulation, you can bring that individual back to life; you can reset his equipment so that if you so desire, he can then fight the next battle at full strength rather than having to fight it at the diminished strength. Usually, at one of these exercises you test the resupply, and you test the casualty evacuation system, and a soldier that is wounded and/or deceased in a battle is taken out of play for twenty-four hours. Then the unit has to go through their individual replacement system to replace him. The observer coach trainer has the capability, if he or she desires, to bring the individual back with the push of a button, to reactivate their system and allow them back into the fight.

*Have you seen this work on a military base in an actual training session?*

Yes, when I was in the service I went through these exercises. Everybody that goes in the box, in the actual training exercise, is always equipped with this. I have used this as a captain, as a battalion commander, and as a lieutenant colonel; I used it going through an exercise as a three-week rotation at the National Training Center in the deserts of Southern California.

*And do the soldiers by and large think they're really in a battle? Is it realistic enough?*

It is extremely realistic and extremely accurate, and consequently if soldiers do not do what they're supposed to do – in other words, if they do not move rapidly, if they do not seek cover – then they will not survive. If they avoid gunfire, they will survive. If they just walk down the middle of the street, they are going to get shot. The system teaches them to operate the way that they are taught to and it emphasizes that if they are covered and concealed, the enemy cannot see them; or if they have objects like trees or buildings or rocks or massive boulders in front of them, then they cannot be shot at by the weapon system; so, your chances of survival are increased. The soldiers, when they don this equipment, they realize they can be shot.

For instance, I'll give you a case in point here in Italy: we started utilizing these systems in trials last summer and the first exercise consisted of a two-day exercise. The first time they went out and they operated the way they normally operated, about 75 per cent of the unit was killed. It was the very first time that they had used the system. They did not know exactly what to expect, and they moved very quickly to get from point A to point B; so we went through it and we explained that if you get hit by the laser, you die. The second time that we went through it, the casualties dropped to 10 per cent, and it took about five times as long, as they were hugging the ground and looking for places to move to. The soldiers in a short period of time realized the validity and the realism of the system.

When you bring commanders into the exercise control room having been on the ground they can see what the soldiers were doing in real life. Then they can go back into the exercise control room and watch that exact same thing in 2D or 3D with the avatars and see the red lines coming out of the avatar (which is the line of fire and whether you hit somebody or not); it gives you a totally different perspective on how to train and the proper ways to train. The realism is wonderful. We are the world leaders in this: we have installed over thirty Combat Training Centers around the world, in the United States, in Canada, in Australia, in Singapore, in Italy, in Croatia, in Lithuania; we are putting one in in Georgia, we have some in the United Kingdom, in Spain and in Romania. This is what we do as our core business and we are extremely good at doing it.

*It's one thing to get shot in a simulated training test; it's another thing to get shot on a real battlefield.*

The one on the real battlefield hurts a lot more. The soldiers learn going through the simulator and there is no doubt in my mind that by training on these systems lots of lives have been saved on the battlefield. Trained soldiers are not being Rambos and GI Joe's: they are doing what they're supposed to be doing and understanding the consequences if they don't do it. Now, this is just in ground mode: I'm sure you have seen the movie *Top Gun*.

*Yes, with Tom Cruise.*

Yes, we did that before the ground phase: we provide a pod, a P5 pod, and an air combat mission instrumentation called ACMI which we put on the aircraft. If you remember, in *Top Gun*, you would have a bad guy and a friendly guy firing during an air fight with these devices (all of which can be recorded also). In that way, the instructor or the commander on the ground, in real time, can see the action that the pilot is taking, and when he lands they bring him into a room and they go over the flying procedure that he utilized and when he fired and the interaction between him and other pilots. From an air perspective, it is also extremely beneficial in training. Obviously, you don't want to be doing that for the first time when you have to do an air-to-air mission for real. Every pilot in the United States (whether they're in the air force or in the navy), if they are flying a fast-moving plane they know who Cubic is, and they have utilized our P5 ACMI system on their aircraft. Even the Marines use the systems to teach their pilots how to fly in a combat situation from air-to-air. It also simulates air-to-ground, so that if there is electronic interference or if there are any defence systems on the ground that are trying to engage the pilots, all the sensors would work so that the pilots would have to evade and escape them. That is also a very powerful system that we use for our armed forces.

*It sounds like an excellent system for training soldiers, because there are so many different layers: there is obviously the digital layer, there are recordings going on at many different levels, and then of course there is the analysis. If you're dealing with 2,000 or 3,000 moving soldiers, it must be an incredible challenge.*

Our lowest operational movement formation is a team, which is usually five people; but we normally talk about a squad which, depending on the country, would be anywhere from nine to eleven people. When you talk about a squad, you can look at the individual soldiers; then you get to a platoon which is about thirty or thirty-five people and that would be three squads. So, you really wouldn't look at the individual actions there; you might look at the icon of that squad and see what that squad is doing. Then you would go to a company level so you have three platoons, and maybe you would monitor platoon movements, and then when you went to a battalion, you would have three to five companies and you would monitor companies, and so forth up from a squad all the way to a brigade, let's say. If you wanted, you could scale right down to individual actions, but you usually deal with the movements and the actions of the unit, being a platoon, a squad or a company. It is a very powerful system: the items I have spoken about thus far are from a live perspective, live meaning soldiers on the ground conducting an exercise and getting feedback. You did mention the data: there are tons of data that are accumulated, and we provide feedback to the training. This is called an after-action review and a take-home package.

*That's just like homework!*

We provide these things so that when they go back and they start developing their training schedules for the next year, they can look at the things they did and may not have done correctly. So, they have this take-home package and they look at the things they did well and they try to develop a training plan to maintain them. And they look at the areas in need of improvement and then develop training cycles and plans on how to ameliorate the things they didn't do so well, so that the next time they go through the learning process, the learning curve increases and the readiness and the capabilities of the unit increase. Even though you've got personnel turnover, if you train to a standard and you know what has to be done, it makes it a lot easier and makes the training a lot better. Data collection is important in order to provide the lessons learnt to help them train better in the future.

This is just the live piece. We also do a virtual piece, which we call the Engagement Skills Trainer: this is a virtual marksmanship judgemental and collective scenario environment tool. What do I mean by virtual? We broadcast a scenario on a screen which is interactive with your weapon. If I want to teach soldiers marksmanship (how to properly fire their weapons, so that when they pull the trigger they are hitting what they're shooting at), we can go into a marksmanship mode. The weapons that we utilize have been adapted so that they have sensors and computer chips in them which allow you to work on the three basic principles of marksmanship: one being breathing, the other being cant and the other being trigger pull. What do I mean by that? When an individual breathes and holds a weapon, the weapon moves up and down.

*Right. Obviously if it's close to your cheek or it's on your shoulder or whatever.*

Correct. As you are breathing, if you don't know how to properly control your breathing while you're trying to aim at a target, you'll miss the target by the simple fact that the weapon moves while you're pulling the trigger. You may be off target: you may be high or low because of the way you are breathing. We have to teach the soldiers how to breathe in rhythm, how to hold their breath before they shoot, so that the weapon is stationary. The second point is cant. Cant refers to the weapon being straight up and down, perpendicular, or at an angle. If it is at an angle (like you see most of the time in the movies), your chances of hitting what you're shooting at decrease. We teach soldiers how to properly hold their weapon upright. And the third that is also very important is trigger pull, because if you don't pull the trigger correctly, you jerk your trigger: as you hit your trigger mechanism, if you do it too fast, that will jerk your weapon, which means it will rise to the left or to the right and you will not be on target.

If you do one of these three things wrong, you may still hit your target. If you do all three poorly, you'd probably be hosed and you are not going to hit your target. Our system provides stoplights on the screen, so that every single time you shoot, it gives you a green, yellow or red stoplight for those three items: it tells you if you are in the proper range. Then an instructor can teach you or work on any faults that you may have in one of the three areas. Now what does this mean for the United States Army? In the United States Army we have nine weeks to transition a civilian to being considered a soldier. One of the items that the individual must do is qualify on their individual weapon.

> *Obviously this virtual training is different according to different weapons. I would assume not all weapons are the same, they don't all fire the same way.*

Your screen is always the same. You can fire different weapons at the screen, you can fire machine guns, you can fire rifles, you can fire pistols – whatever you want – but the mechanism between you and the screen, the laser, is the same. Obviously, if you shoot a machine gun that is firing three rounds every second, you are going to get more impacts on the screen. You are still going to have to aim and there is going to appear a larger burst area, but if you are not aiming or shooting properly, you are not going to hit your target. If you are shooting a machine gun or if you are shooting a rifle or a pistol, it doesn't matter. The marksmanship aspects of it still have to pertain and our system allows you to do both. In marksmanship mode, for three seconds before you have point of impact and for five seconds afterwards, we can track the performance by different colours. It will be yellow before you'll have a red dot, and then it will be green afterwards. You can see how your weapon is moving before you physically pull the trigger, so you can see if you're on target or if you're all over the map as we say. These are rules that increase the individual soldier's capability of engaging his target.

One of the things you have to do in order to qualify as a soldier is pass this test in a weapon system. By putting trainees on our Engagement Skills Trainer, when soldiers go to the range the first time to fire their weapons, 96 per cent qualify after their first go. This has created huge time-savings: I have heard somewhere in the neighbourhood of $50 million saved in ammunition, because you don't have to go back to the range – you save in time, effort and so forth. The Engagement Skills Trainer has been extremely beneficial. We have been selling them for years to the military, and it teaches soldiers how to do marksmanship. From that we have gone on to 'scenarios', and a scenario is like a movie. You can create whatever type of scenario you want to: an offensive scenario, a defensive scenario, you can create checkpoint scenarios – anything that you want to film can be created in this format and then you can bring multiple soldiers on line. You can bring up to a squad, so now that squad leader can assign sectors of fire. He can assign portions of the environment – who shoots whom, depending on what's going on – and you can teach at the squad level a lot of things that you can do in a virtual environment and then later go live. All of this increases tremendously the capabilities of those individuals and of the units. Commands can be given to the soldiers as normally would happen in a real environment.

The final thing that can be done in this virtual world is adapted to police forces and it is the judgement call, or what we term 'shoot-don't-shoot'. You can imagine a hostage situation (like a bank robbery): an officer walks in and sees a bad guy on the one side and, on another side, other people and civilians. There are all sorts of things happening and an individual holds something up in his right hand and you have seconds to decide: Is it a pistol? Is it a phone? Is that a baby?

*I'll bet you don't have seconds but fractions of a second.*

Exactly. What you need to do is train the eye or train the reaction of the individual: is it a friendly or is it a threat? And the police officer, as he comes in, is providing information and instructions, like 'This is the police. Drop your weapon!' and as things develop, he has to decide: does he shoot or does he not shoot? Is this a threat to him or is it a telephone?

*That could make the difference between life and death or continuing a career or stopping one.*

From the perspective of a police force and also from a military standpoint, that is true. Often, as you enter a room to clear it, you could have people there and your rules of engagement might differ for the size of the unit in training. In our state-of-the-art methodologies, you have realistic content and it allows you to train across the full spectrum of all types of conflict. We are able to interoperate between the various modes (live, virtual and constructive). The soldiers that are on the battlefield, you can see them on the virtual screen and they could also be seen in a constructive mode (on a computer) so that way you don't necessarily need to take large units in the field in order to be able to train different units. You can put your staffs in a constructive mode and then have smaller units out in the field that are actually manoeuvring. You can put some in a virtual mode and yet you, as commander, would know what environment they were in – whichever it might be – because you could see as big or as small a picture as you wanted.

*Yes, that makes perfect sense.*

That is the beauty of the technologies that have changed the way we train in the army now.

*Can you give me an update on what Cubic does in terms of transportation?*

Yes. Cubic Transportation is the leading integrator of pavement and information systems. Their motto is: 'Intelligent travel made real.' They provide intelligent applications for travellers and traffic management and deliver tools to travel and to make decisions concerning the smartest and the easiest way for them to pay for their movements and enable authorities to manage high demand in real time.

Take for example the Tube in England which is their hallmark: they have created the Oyster Card. An Oyster Card is similar to a credit card because you can put money on it and when you touch one of the devices at the entrance of the subway, it knows that you entered at that point. When you exit, you tap it again and it knows where you exit and it calculates the cost of transportation and it deducts that from the credit that you have on your Oyster Card. Rather than going in and having to say, 'I want to go from point A to point B', and then the system says 'It's three pounds fifty' and you have to put the money in and then it prints the tickets. All of that is still possible but, with the Oyster Card, you can put £100 on it and every time you tap it on a device, it deducts the correct amount and you know how much money you have left.

Now what they are working towards is cell phone technology: your cell phone is linked to your bank and your credit card. For instance, if you agree to be in their intelligence system, you will be informed by your telephone when you get off at your Tube stop and you usually catch some line that may be very congested that day. So, when you get off, the system suggests that you take a certain bus because it will get you to where you need to go quicker. All of this is provided in real time to the individual on his mobile device. Oh, by the way, the system will say, 'Here's a twenty-per-cent-off coupon if you want to stop at Starbucks', because that would be very near the bus station: while you wait for the bus, you can partake in a cup of coffee. The system will do it all through your phone.

*Is there any human involved in that exchange of information?*

There are obviously humans behind the scenes, but it's the software solutions that we have created that have allowed this interchange between individuals and the traveller. As I mentioned before in terms of data and situational awareness, we have a data analysis company and they look at London or they look at Washington or any of the cities where we have a system. They will examine the flow of transportation and then go to the transportation authorities and tell them: 'You know, you've got four buses going on this line, but you don't have anything going this way. If you moved one of your bus lines this way, it would improve traffic over here.' It is a phenomenal way of providing feedback to the customer or to the clients on how to manage their demand for transportation and free up (especially during high peak times) and make the transportation better for them and obviously more comfortable for the user.

*The sheer numbers involved can be staggering.*

We are talking billions of dollars and millions of people that keep using these systems. I don't have the figures directly in front of me.

*In any case, that is a lot of people. You are really giving a new concept to what a smartphone can do, in the sense that my phone is suggesting better routes, it's giving me a coupon for Starbucks . . . Are you concerned about privacy?*

I'm not 100 per cent sure on this because it's not in my specific area of expertise, but I know that there are very strong firewalls in place which protect the information. Also, I know that individuals have to agree to be in the system, and agree to the passing of information. If all you want to do is be able to pay from your phone, you can do that as well. If you want to be in the system and provide it with information, you have to agree to do that. In terms of all the security and possible hacking and everything that's involved: these are secure lines, so nobody is going to break into your bank account. It is secured and it is highly insured. It's just like when you go online: you have so many benefits. We've been doing this for many years, and the London Transit Authority and Cubic are joined at the hip. The accolades that they have for this is phenomenal and it is a fair collection system: we collect the fares for them through this mode, through the touchpads or through the technologies in the smartphones or in the smartcards.

*So, you do recognize cyber security as an issue that needs to be addressed?*

Certainly, certainly – it is an issue, and obviously the system would not be beneficial if it could be hacked. The security on all of these things even in our system is tight: they all operate within certain frequencies. The country where we install the system provides us with the frequency ranges that we can use, so that when we set our devices, they are set within those frequencies. Therefore, the radios or televisions or cell phones don't interfere with what's going on. From a cyber security perspective, we have to take that into account and it is always a major factor in what we do.

*And you work on a global scale. Do you come across difficulties of dealing with perhaps countries that are not very friendly to the United States? This is obviously expensive technology and extremely useful. I don't think you'd want it to get into the wrong hands.*

In the United States, we have a lot of regulations concerning whom we can sell to and whom we cannot sell to. Before we can sell to anybody, we have to get the proper export licences and so forth. The company has been in existence for over sixty-five years, and it is traded on the New York stock exchange. All the rules and regulations, the Department of State and the Department of Commerce are taken into consideration. Obviously, there are countries in the world to which we are not allowed to sell, and you can imagine which ones those are.

*The funny thing is those countries can change: so your company has to be updated constantly.*

But the countries to which we do sell all have provided licences for us to be able to export to them. And most of them are allies. We obviously don't sell to enemies of the United States. We have our systems in North America; we have them in South America, in Europe and the Middle East. We are in the Far East in Asia, we are in Australia . . . We are global leaders in these areas.

*What do you foresee in the future, let's say five to ten years down the line? Not just more of the same, because obviously your products are going to become better at what they do. Do you see a gradual improvement? Do you see something else coming along in that time period?*

If I knew about Google, Facebook or Apple before these things came out and had invested in them, my life would probably be a lot different today.

*You would have your own private jets! You wouldn't have to work.*

I can tell you in terms of global defence that our president is looking at some items: he calls them 'Next Mission' and 'Next Training'. What does 'Next Mission' mean? In the United States (and probably in a lot of other countries), we always have the tendency to train for the *last* conflict. So we were training for the big wars as if they were to take place in the desert, and when more recent asymmetric wars came up with huge numbers of civilians on the battlefield, with improvised explosive devices, it shook up the way we think about warfare and it took us a while to figure it out. What we want is to be trusted partners and to be think tanks: to look and see what the future will in fact bring, what the next missions will be in which our country or our allies may be involved. Then, we want to develop training systems and simulations and technology that can provide an advantage for those new missions. We always say that you never want to go into a conflict without an advantage; in other words, you never want it to be a fair fight. You want to have overarching dominance and advantage over the enemy in terms of intelligence, situational awareness, firepower, manoeuvring, command . . . everything to make it an unfair fight.

It is a very interesting historical period and yours is a very good question. We are looking at the future and focusing on the tendencies we see now. We are working on developing smaller, lighter, cheaper, longer-lasting batteries, integrating them into the systems that the soldiers carry so that they do not need additional batteries. We would like to develop a weapon that already has a laser engagement system on it; perhaps a vest that already has the sensors in it; maybe a radio that already has the exercise control capabilities or the frequencies in it so that you don't have to wear additional kit – it's embedded in the stuff that you already have. These might be possible solutions. Do you know where cell phones will be in five years? You see movies nowadays where they are talking about implanting chips in your hand so you don't need a device. You put your hand to your ear and talk that way.

*It sounds like science fiction but it actually is possible.*

The things we are doing now were science fiction fifteen or twenty years ago. Whoever would have thought of some of the things that we've got now? It's definitely something that we're looking at as an organization. Cubic Global Defense in its current strategy for the next five years wants to develop 'Next Mission' and 'Next Training': two major items that our president wants to focus on. You can't train to the last mission, you have to train to the next one.

*Exactly. Now you have a very special perspective because you've been around this stuff for a long time and you saw the revolution happen in terms of technology. How were you able to embrace it and understand the technology, as the change came so fast? I'm speaking about the last thirty years.*

When I first came into the army, there was a system where you would have a number on your helmet, and the individual would say, 'Number 58: I shot you!' And you would say: 'I'm not 58, I'm 68'; and then with time and laser technology with scopes, MILES came into effect. Before, the gear they wore was very big, but now it's gotten smaller. It's similar to the theme in the movie, *Field of Dreams*: 'Build it and they will come'. Or, the proof is in the pudding. As these systems were developed and we saw the benefits, we had no choice but to embrace them. Now, I grew up with computers, but I certainly cannot do stuff with computers that kids today can. I can do FaceTime on my phone; I know how to do apps; I can talk and get email; I can send messages and create videos. But some of the things that the kids do today are truly spectacular.

What have we done in the military? For example, twenty years ago when we developed our systems inside of tanks, we knew that the kids coming in were experts at using PlayStations or Xboxes. The devices inside some of our field vehicles were developed to be similar to the devices these kids were used to operating. Our helicopter pilots, for instance, have a drop-down optic device that goes in front of their eye: that would be similar to Google Glasses. I could not fathom looking in front of me and seeing something in my right eye, being able to fly the helicopter or shoot a weapon, but for kids today this is second nature. For us dinosaurs, we have to learn how to do something; but today's kids spend most of their lives doing this kind of thing. In some cases, they don't know how to talk or write; they'll be sitting across from each other texting rather than conversing, but that's the technology. The leaders of today hopefully have the vision to see the power of these devices, and their benefits, such as the increase of readiness, the decrease of costs and resources, and the time that is being saved. I think it's safe to say that leaders have no choice but to invest in new technology.

The point that you brought out is important: is it secure or is it safe? Look at the unmanned vehicles or look at where we are going with robotics. We've got robots now that are going out and dismantling bombs. If a robot is blown up, it's just a piece of metal – it's not a human being. The individual who's flying that unmanned aerial vehicle or who's manoeuvring that robot is behind the computer screen and telling that thing what to do. We've seen this in movies. In the future, those robots may have some form of computer brain that allows them to operate differently than today. Only the imagination can tell you where we are heading with these technologies. Just think, that you can take your document or your book that you're going to write and you could put it into Google Translate and in about ten seconds it can translate everything from English to Italian or French or Chinese . . . the power that these devices have is scary.

*You are talking to someone who has also embraced the technology.*

From a training perspective, we increase situational awareness whether it be for transportation purposes, for soldiers, for sailors, for airmen or for policemen: we want to help them do their jobs better and, if they are deployed, to return home safe and sound. That could be from walking the streets, being on the beat, to being deployed to a combat theatre: we increase their situational awareness, their readiness and their capability.

*And a final question that I always ask is this: what advice can you give us? You have already mentioned some things in this area but from your perspective you probably see all kinds of reactions: people who are scared; some may be apprehensive; others are gung-ho and maybe too enthusiastic. From your perspective, what would your advice be?*

I've been out of the army for nine years now, but I was in the army for thirty-five years. I can speak as a former soldier and as a retired officer that spent most of his time in the operational field, training soldiers. I would say that you have to have trust and confidence in the systems that are provided. Case in point: I jumped out of aeroplanes most of my life. The parachute that I wore every single time I jumped was packed by an individual called a 'rigger' and I had to have trust and confidence in the system that that chute was packed properly, that it was inspected; and when I jumped out of that aeroplane going 125 knots at an altitude from 800 feet up to 24,000 feet (because I was also in special forces and did some freefalling), I had to have confidence that my parachute was going to open. I knew I was going to land safely because I was properly trained. I knew how to do a proper parachute landing fall, so that I would not injure myself and then I would be able to continue on with the mission. So, my recommendation or my advice to your readers is to have trust and confidence in these training devices that we provide to the force, be they virtual or constructive or live, and take them for what they are: they are tools to help improve situational awareness and increase readiness. The situations will change, but hopefully by using these devices, there will be no surprises when soldiers go into live combat, because they've trained well. That, in itself, has a huge impact on the psyche and on the confidence the soldiers have going into whatever situation they may face.

Cubic Global Defense is enabling a safer world. In terms of transportation, it provides intelligent travel made real. Our ideals have been thought out for what it is that we do for our clients and that is what we're all about: satisfying our customer is our number-one priority, and we want to be their lifelong and trusted partners so that they continue to come back to us. We increase our own capabilities by listening to our customers' feedback. We want to provide them with the necessary tools to enable their soldiers to carry out their missions properly and to return home to their loved ones.

*And the statistics speak for themselves in terms of the efficacy of these devices.*

Soldiers need to avoid taking gambles, but they do have to take educated risks at times, as they may have done in the past. Following the tactics, procedures and doctrine at the unit level makes you better overall. Training is the cornerstone of our force. Soldiers have to be trained and ready and so the best training is what we're all about: providing our customers with the best training tools for them. Our motto at Cubic Corporation is: 'Global, innovative and trusted'.

*That sums it up.*

## 5.
# *Social Media*
## *(Breanna Fulton)*

At lunch in Rome with some friends and their daughter, Breanna, we ended up talking about how digital technology is changing society. While discussing some of the people I had interviewed for *Connected World*, Breanna (who was fourteen at the time) pointed out that if I really wanted to understand the full impact of the digital revolution, as well as the thought-leaders driving technological developments, I should also interview the people most likely to be engaging with new technologies on a day-to-day basis: so-called digital natives. And how right she was – the conversation that follows, with Breanna and her friend Lily, neatly illustrates how it is not just those at the forefront of their industries who are leading the digital revolution, but also those for whom disruptive technologies are just a fact of their daily lives.

* * *

*Breanna, explain to me the difference between Instagram and Facebook. Why are people your age leaving Facebook to go to Instagram?*

Well, it's because Facebook has become attractive to a lot of older people [chuckles]; I'm sure young people don't want their parents following them. A lot of people just block their parents without telling them, because they're rude. And then Facebook, it's just growing old. It's been there for a while so they want new things, and Facebook is like a combination of Instagram, Twitter and other things because it has all the stuff that they have, but these just pick them out individually. Instagram is like the photos of Facebook.

*But you have a lot more than just photos. Look, you have followers, you have comments, you tag people.*

On Instagram, all you can do is post photos. You can't post your status, you can't comment on someone's page. I might want to comment on my friend's page (her name is Amy). So, I can search for her here. But if I wanted to comment on her page, I can't. The only thing I can do is make a comment on one of her photos. So, like, she posted a photo of the Color Run. I could say, 'Oh, that's really cool.'

*And do you do that?*

Yeah, I do. Girls are always nice to each other on Instagram, or 'fake' mean to each other. Somebody would post a selfie, and then they'll be like, 'Oh, you're so pretty.' But then you'll get the occasional 'Eww' as a joke, but you're still best friends [laughs].

*Yeah, it's kind of complicated. I don't think I'm there yet.*

You're right: we're kind of weird.

*Do you want to talk about Tumblr or do you prefer not to?*

I don't have a Tumblr account, so I can't talk much about Tumblr.

LILY: I started a Tumblr account just for a blog. I'm still not fully understanding it, but I just started it for my blog for travel. Tumblr is very popular, though.

BREANNA: It used to be.

LILY: It's very popular now.

*It is or is it not?*

LILY: It is. Tumblr's like the new thing.

*What is it?*

LILY: People, I think, had Tumblrs back in middle school. I remember Tumblr and Pinterest were super popular a couple years ago, and they have faded off since. Pinterest is, like, gone off, but Tumblr is now really big.

*Is it a dating site?*

LILY: It's big in the United States, too, because of YouTubers now. Tons of YouTubers have Tumblr. I don't really understand. Tumblr, I think, is a more interactive thing, like between comments. In Tumblr, you can have this stream of comments, like different people commenting on one thing, and it seems a lot like Facebook. You can post words or pictures. I used it because my blog is through Tumblr.

BREANNA: It's so uncensored, though. Tumblr is pretty bad.

*Yeah. There's some pretty unsavoury things, I think, that happen there.*

BREANNA: No, you might mean Tinder.

*Oh, Tinder? Sorry, what's Tinder?*

BREANNA: Tinder is a dating site. It's a really shallow dating site. 'Dating' is not the word for it.

*It's a hook-up site, isn't it?*

BREANNA: Yeah, essentially.

*Tinder uses your cell phone's locating service in order to find anonymous sex.*

LILY: I have Tinder. I'm not going to lie to you, I have Tinder. That means some people will write on their Tinder: 'Looking for a fun night', you know what I mean?

*Right.*

LILY: Some people are like that. Other people (like on mine now) will write, 'In London, looking for some friends to show me around or see the city', or something like that. It does use your location.

*That seems a little bit tamer than, 'I'm up for whatever'.*

LILY: Yeah, it uses your location. The concept of it is very shallow because you literally just see pictures of people.

*Have you ever had a bad experience with it?*

LILY: With Tinder? Yes. In the States. I've been on dates from Tinder with real people, and yeah, it is like getting hooked up; it is literally like hooking up with them on a date. Not like 'hook-up' exactly, because it could turn into something more [chuckles] but it's more like a connection, I would say. Some people definitely use it for the wrong reasons. But it's solely based on your appearance, that's kind of the sad part of it. You have your pictures and you have a max word limit. You can swipe 'nope' or 'yes' if you like them or not. And if they match you, then you can message from there.

Realistically, if you walk up on your college campus, and see a guy, the first thing you're going to say is not, 'Oh wow! He's really got a great personality' [sarcastically]. Really what you're going to say is, 'Wow, he's hot.'

*Go ahead, Breanna. This is like a parenthesis, but it's interesting. But this is your whole point, Breanna, you're saying, 'I need to know what these people are experiencing and where they're going.'*

Well, this is the new generation. If you were opposed to all this, you would basically be disconnected. This is my blog with all the stuff I've posted. I have 135 followers and I'm on a private account, so nobody can just follow me if they want to. But they can request to follow me, and then I can accept it.

*But this is like Facebook? It's like friending someone on Facebook.*

No, because you don't have a status, you can't directly talk to them, and you can only post photos.

*For example, I'm friends with my sister on Facebook and she has a lot of stuff that she posts on Instagram. I see it all through my Facebook account with her.*

Going back to Tinder, they'll probably try it with everyone they match with and then say, 'Oh, someone's into this. Here we go.' It's like playing the Lottery. Somebody's going to win. Somebody is going to say 'yes' eventually because you know someone out there is looking for the same exact thing. And it's very regulated.

*So how did you know about Tinder?*

Because so many of my friends have it.

*Have you asked your dad if you could have one of those accounts?*

Tinder? Oh, no. If you would like me to talk to you about YouTube, I can. Okay, so on YouTube, you can post any videos and people will watch them. Now, there's a new generation called YouTubers, and YouTubers are people who vlog on YouTube. There are a bunch of twenty-year-olds, who sit at home with a camera, and they talk to it, and they have over a million subscribers. So many of the famous YouTubers are now writing books. They're all authors of their own books. It's about all the funny stories they have, and they are so popular – everyone wants to read them. There's this girl called

Zoella. I don't particularly like her. I didn't think she was that interesting. She is one of the most famous YouTubers. She has ads for all of her YouTube videos. Now, she's advertising beauty products. What has she done on YouTube? She would tell funny stories. She'd do collaborations with other YouTubers. She would do make-up tutorials, and she has tons of 'likes', she has tons of ads, and she wrote a book, and people could pre-buy it. It sold out quickly, and she sold more copies in a certain amount of time than J. K. Rowling did.

*No!*

She outsold J. K. Rowling in the first day –

*That's amazing.*

And this is a book that she wrote because of YouTube [laughs] and just sitting in her house. Her book outsold Harry Potter in *x* amount of time. And she's not even an author, but she decided to write. She wrote a story, and then other people, they're now all writing stories. Everyone's pre-ordering these books. They're all really excited for them. So now all YouTubers are turning into authors and business owners.

*Real pages, which we thought were obsolete?*

And YouTubers get paid by YouTube. They get paid, actually, a pretty fair amount of money. They're twenty-year-olds just staring at their screens but people are fascinated by them. The YouTubers go to all these conventions – there are YouTuber conventions – and people will come to see their YouTubers.

*Isn't it funny that these people who started with digital formats are now writing books? Which you would think nobody's doing any more.*

Yes. Plus, they sell merchandise. They each have their own product-selling websites, and you can buy mugs with their names on them, you can buy signed posters. They were originally supposed to seem like just any other person. They're not celebrities, so you can really begin to relate to them, and they're really funny. A lot of them are really funny.

*Do you think this is the new generation of celebrities?*

Yeah.

*Like the Brad Pitts of tomorrow?*

People aren't watching TV any more. People are like, 'Oh, what's television?' People watch YouTube and Netflix, and illegally stream movies that are in theatres. A lot of people don't watch TV any more.

*But you mentioned that there are popular TV shows and TV episodes.*

There are popular TV shows and a lot of people watch them, but a lot of TV-watching now is on Netflix and such. I will go on to YouTube, even me, and I'll be like, 'Oh, I just want to watch one video.' Three hours later, I'll be like, 'Oh my gosh, I thought I'd only been watching four videos. It's been three hours.' That's how time-consuming YouTube is.

*What's that called?*

Binge-watching [laughs]. Binge-watching is when you don't watch something for a while and then you watch it all at once. I'm really good at doing that with YouTube because I don't watch YouTube very regularly. I have this one friend who is so overly attached to her YouTubers that she follows. She will see that they have tweeted on Twitter that they are about to post a new video, then she will go into YouTube, wait for their video to be posted and get one of the first 300 views, comment, and see if they respond to her comment because it's so quick. She's hoping to get responded to.

*But why does she have to be one of the first 300?*

She'll be one of the first 300 people to watch it, within, like, five seconds from it being posted, and she'll comment how good it was, and sometimes they'll respond back. It's known as so good for them to respond back. It's like, 'Oh my gosh, they responded back to me.' So, it's more of a celebrity status thing.

*Which means you ranked, right?*

Yeah. It's like, 'Oh, yay!' And somebody even noticed she was doing that so often to their account, and he wrote: 'Oh, I'll thank you, you're always there.' And it made her so happy. She loves her YouTubers, she loves everything they say. She watches YouTube so much.

*What's your reaction to that? Would you do something like that?*

I will confess, I do watch YouTube. Not as often as a lot of people. Some people are done with YouTube, some people are just overly obsessed. I will go into YouTube and there's this one YouTuber I really like because he's so funny. Because, in reality, he used to be a really awkward guy. Now, that's really changed through his video-making, and he collaborates with other people and it's hilarious. I love watching it, but he barely posts. Because of his YouTube channel, he has now got YouTube to do tons of stuff for him and his friend that he lives with. He has gaming channels with his friend. Because of YouTube, his talent was discovered, and now he and his friend are on BBC radio every single week.

*You're kidding.*

They have a job with BBC radio, every single week.

*So what does that tell you?*

It's a good way to get found.

*But the BBC is still over and above YouTube, it seems to me.*

It is, but maybe only right now. More of my generation is watching YouTube. And things like talk-show hosts? Jimmy Fallon and Conan and all them?

*Yeah.*

They're on TV, right?

*Right.*

Yeah. All the teenagers, they still watch that. But they don't watch it on TV. It gets posted to YouTube. So now, TV shows and TV interviews are on YouTube. And they have tons of subscribers. Jimmy Fallon has millions of subscribers, he has millions of views!

*And who owns YouTube?*

I don't know.

*It's a rhetorical question, because of course Google owns YouTube.*

Does it?

*Yep. Google purchased YouTube in 2006 for $1.65 billion.*

People are using YouTube such a lot. These YouTubers, a lot of them had problems and they share those. They talk about their personal lives with millions of people watching. There is no such thing as a private YouTube. You *can* have a private YouTube account but it's rare. You don't go looking for private YouTube accounts. In their description after their video, they'll advertise all their other videos, and you can go to their shop and buy their merchandise. I have a YouTuber top, and it's from the guys I was talking about, Dan and

Phil. One of their quotes is on the shirt. People buy that stuff and they love it.

*Do you think YouTube is heavily regulated?*

It is. But could it be more? Because on my YouTube, there are certain videos that I can't watch because my Google account knows I'm under eighteen [laughs]. Since I use it through my email, it knows I'm under eighteen, and it says, 'You must be eighteen to watch this.'

*Can you lie?*

Well, I would have to lie through my email. I started my email when I was ten, and I said that I was thirteen; so now, YouTube thinks I am seventeen (but I'm only fourteen).

*Couldn't you just create another one?*

Yeah, I could if I wanted to. Tons of my friends just put down that they're thirty. Because you have to be thirteen to make an email account, so they lie about their ages.

*But if you lie, what happens?*

It just makes you older. On their YouTube account, they're over eighteen, so they can watch those videos and it's fine. Some videos are for people over eighteen, but the site doesn't actually know your age. There are videos that don't necessarily show any images, but still discuss them. There will be some YouTubers who are posting videos of themselves watching weird videos (like pornography, for example). They don't show anything provocative but they show their reactions to it. I could watch it if I wanted to (but I usually don't watch that kind of video). They all put neat titles to these videos to make you want to watch them. It's like, 'That one was watching strange porn.'

People do Q and A, so other people post their questions on Twitter and then they answer the questions in their videos. One of the questions might be: 'When was your first date?' They'd answer quickly and then they'd move on. But the title of the video would be *My First Date*. So you would be like, 'Oh, I want to hear about his first date', and you'd watch it. And then he mentions it but maybe it's just along with a lot of other questions and stuff.

*It sounds superficial to me.*

It's really interesting to listen to other people's lives. Then, at the same time, it is a little weird. But technically that is what TV is doing: except that on TV they deal with fake lives. For YouTubers, the lives are real, so it makes it more relatable.

*Is it?*

They talk about their awkward moments, even though they are obviously hiding some other moments. There are some people who actually *do* talk about stuff, like Zoella. She has anxiety issues, and she shares that with YouTube, she talks about that.

*It's kind of like a permanent reality show, isn't it?*

Yeah. Except they're talking, and you feel like you know them. The amount of time people spend watching these videos is amazing. For example, I'll watch a YouTube video posted two years ago, and then I'll see people who talked about it two days ago in the comments, and they're like, 'Oh, "like" this comment if you're watching this in 2015 even though it was posted in 2013.' And you'll see maybe 100 likes. So, this video is three years old and it may be forgotten about but not for the thousands of people who are still watching it.

*I don't know. You're not really obsessed with this, I mean, you don't spend hours on it, do you?*

I can get hooked on it. When you're watching a video, on the sidebar, it'll be suggesting videos for you to watch. You're like, 'Oh, that has a really cool video, I'll watch it.' But the way YouTube deceives people into watching more videos is that they say every video is only five minutes. Even if the video is longer, like twenty minutes, people still say, 'Oh, twenty is enough, it's okay, I can do it.' It goes by so fast. Some videos are twenty minutes, but that's a long YouTube video. Usually, they're three to seven minutes, so you're like, 'Oh, it's only another five minutes. I can do that.' But then you keep repeating that to yourself and you spend four hours on YouTube, when you weren't meaning to, you just wanted to watch one video before bed.

*Yeah, but that's just successive superficiality. So you've been there four hours but you haven't really digested four hours of video, you just digested little pieces.*

What seems weird is they post YouTube videos of people reacting to other YouTube videos. There is a whole channel called Teens React, and it's teens reacting to videos. And it's funny to see their reactions.

*I don't know if I would find that funny, but I guess people do.*

I watched a few. They're actually kind of interesting, but only if they have bold people in them. It has to be partly put on, because if they were watching that alone, they wouldn't be speaking out loud and laughing out loud. But since they're being filmed, they're like, 'Oh, that's cool!'

*So it's staged?*

Yeah. Well, it's not completely staged, but they're saying everything they're thinking out loud and a little bit more. If I were just to film myself, watching a video, then it –

*It would be boring.*

It wouldn't be interesting.

*No, absolutely not.*

A lot of teenagers my age are on the internet all the time, and it's causing a major issue today because people are spending so much time looking at porn and other things. People wouldn't have the idea to do that if they didn't know about this. How do you think they're finding out about it? I was watching this YouTube video, and it was about these guys talking about what they like in girls [laughs].

*I would imagine that's obviously pretty popular.*

I was watching it because I knew people said it was really, really sexist. I was watching it just to see. The video showed them saying, 'We like girls who are tall; we like girls who can sing.' People are really outraged because they're like, 'You're saying all guys only like girls who can sing. What if you are a girl that can't sing or can't do the other things you just listed?' That just makes girls more self-conscious, because a lot of girls don't realize that's just what a couple of guys think of girls. They think that *all* guys are like that, so it makes them self-conscious.

The youngest girl watching that video claimed in the comments that she was seven years old. There are people in those comments asking for relationship advice, just asking anyone. I saw these two comments: it was this girl saying something about her boyfriend. Then there was this other girl who says, 'I don't know if I can help you, because I am eleven, but I've had two boyfriends, so here's my opinion on it.' I was like, 'You're an eleven-year-old watching YouTube videos giving somebody advice about dating?'

*On relationships.*

I would never ask a public question like that. I wouldn't ask perfect strangers for advice over the internet.

> *I'm just not sure that when you're looking at comments, you're really dealing with the people as they say they are. I don't think you really know. It could be a 75-year-old man posing as an eleven-year-old girl.*

But what's the point in giving dating advice? That's what people have come to. What's the point in that?

> *Yeah, I think people need to just leave YouTube for a second and meet each other face to face.*
>
> *Can we talk about self-harm for a second?*

Self-harm can be common among groups of adolescents and a lot of the time this can relate to what is on the internet and social media.

> *Do you have any experience with this happening with people you know?*

Yes. Last year I knew a girl, let's call her Sarah, who had previously self-harmed for other reasons and Sarah searched the word *purge* on Tumblr to see what would come up, and horrifying images of cuts people had given themselves showed up, all posted on social media. It's really bad to see how public these images were because many depressed teenagers wouldn't even have the idea to harm themselves if they didn't know that it was a thing, or hear about it on social media. Also, social media can be a cause of self-harm because a lot of girls and boys post pictures of their perfect bodies and all of the social events they go to. This makes other people feel badly about themselves to the point where sometimes they begin to hate themselves.

But that's the thing about social media, it can be good in some circumstances like when you see cool ideas, funny videos/images or what your friends are up to, but other times it can get really horrible. I often get really annoyed at social media posts when I see girls at my school posting bikini pictures on Instagram or pictures of cool parties that they have been to recently. It can also be worse on Snapchat. To post a picture or video on Snapchat, you have to take the picture/video right as you are posting it, and it only stays on your 'story' for twenty-four hours. Because of this, people can get away with posting stuff that they wouldn't get away with if they posted it on Instagram or Facebook. The sort of pictures I'm talking about are tons of selfies, pictures of parties and hang-outs that a lot of people weren't invited to and gym pictures of topless guys and revealing selfies of girls. These types of photos just make others feel worse about their bodies, or annoyed that they weren't invited to parties that their friends were at and in turn can help cause some depression or self-hate.

*Why do these people self-harm? What does it solve?*

It's hard to fully understand when you haven't been in that place. I have a hard time understanding it sometimes as well, but I have had multiple friends go through it and have helped them. The idea is that they hate themselves so much that it makes them want to hurt themselves, but as I said before, I believe that most people wouldn't have the idea to harm themselves if they didn't know that it was a thing via others and the internet.

*Okay, thanks, Breanna. It's been great conversing with you.*

# 6.
# *Advertising*
## *(Sir Martin Sorrell)*

Like most people, I had never heard of a company called WPP, let alone understood what the letters represented. None the less, a mutual acquaintance suggested that I speak with its founder and CEO, Sir Martin Sorrell, to pick his brains about the new digital era, and so we were introduced. And believe me, nothing can prepare a Catholic priest who teaches philosophy for an interview with Sir Martin.

WPP is – or so I discovered – the largest advertising company on the planet, and Sir Martin is a constant feature in the global media as one of the most influential Britons in the world today. Although based in London, his is a global enterprise with operations around the world. In our conversation, we touched on many diverse topics. What follows is a sampling of those. At times surprising, but always entertaining, our dialogue was an eye-opening opportunity to get Sir Martin's views on a wide range of the crucial issues of our digital age.

* * *

*I'm interviewing people about the impact digital technology is having on their sector of society. I noticed that one of the areas that I did not touch on in the first book was advertising or publicity. So I said, 'Who do you think is the best person to speak with in this sector?' I was told, 'Sir Martin Sorrell, who lives in London.'*

Okay, good enough.

*So there you have it. I've been told that Sir Martin Sorrell has changed the advertising industry in the last thirty years.*

I would think that's an exaggeration. My mother would have said that – if she was alive, she would say that.

*Why do you think people would say that?*

Well, first of all, I was in Saatchi before I started WPP. I started WPP thirty years ago when I was forty, and for the previous, well, I guess eight, nine years I was CFO – chief financial officer – of Saatchi and we built that with the Saatchi brothers into the largest company in the world, in that sector. I guess we've done it again but this time with WPP. WPP stands for Wire and Plastic Products. It was a small engineering company quoted on the stock exchange. I bought control of it with a stockbroker and it had a market capitalization of £1 million; today it has a market capitalization of £19 billion, $30 billion, and we have 200,000 people in one way or another in 112 countries working for us. The 112th is Cuba, the 113th (if sanctions come off), will be Iran, although as a Jew I'm not particularly happy about that for the obvious reason that Iran seems to be dedicated to destroying the state of Israel.

*They have said so.*

They continue to say so, yeah.

*I'm not sure if that's propaganda.*

Well, at least one hopes it's propaganda, but there are people who worry about that. Iranians are super-sophisticated people. There's a literacy rate there of about 99.5 per cent, both men and women, which is quite extraordinary.

*It is.*

It's a third bigger than Myanmar, which we went into, that was the 111th country. So, I guess people look at that and they say, 'Well, he was involved with Saatchi; he's involved with this,' and certainly in both cases we started from very small roots and very small beginnings. I've always been writing that annual report. I've mimicked deliberately what Warren Buffett does every year in the Berkshire Hathaway report. Norman Pearlstine, who is Chief Content Officer at Time Inc., wrote about my annual report recently in a blog saying, 'It's probably the best annual report in the world, better even than Warren Buffett's.' So, I know Warren a little bit. I wrote to Warren and I said, 'Warren, Norman says this is a better annual report even than yours.' He wrote back – he's a real class act – he wrote back and he said, 'There's nobody in the world that I would rather be second to than you [laughs].'

*That's wonderful. That is wonderful.*

So, that's another letter I have on my mantelpiece. I think people look at that.

There are two buckets really: one is the geographical bucket, and one is the technological bucket. We spotted the rise of China. I started WPP in 1985, and I decided to have our first board meeting in China with WPP in 1987–8. We had one this year in Beijing. We generally have them in London or New York, but we have one board meeting outside the UK and US, in Europe, Asia, Africa or Latin America. We've done probably four or five board meetings in China over the years.

We spotted the rise of the BRICs: Brazil, Russia, India, China; and the 'Next 11', meaning Indonesia, Nigeria, Mexico, etc. The other thing we did was we sort of spotted the technological changes, and technology, data and content have become important as a differentiator and in areas that our clients are interested in. If anybody does say what you said – apart from my mother [laughs] – I think it was because we flagged those changes, and we've also flagged the need to integrate our offer. Our strategic objectives are four-fold. The first I call horizontality, which is the integration of what we offer: getting our people to work together for the benefit of clients now is at the top of the agenda. Number two is the fast-growth markets, which are 31 per cent of our business and we want that to become 40 to 45 per cent. That is what 'BRICs and Next 11' means. Digital is 37 per cent of our business, and I also want that to be 40 to 45 per cent. And finally, data – we've achieved that objective, in that measurable marketing services represent 50 per cent of our business and that's where it should be.

*I'm not sure what data means.*

It means statistics. It means we have profiles. We have profiles that we own, anonymized profiles of everybody who lives in the United States, all 300 million of them. We don't know it's you, but we know it's a number or a letter.

*That is scary.*

Well, it is scary at one level.

*Because someone does know the name.*

Yeah, but we don't.

*Right.*

It's anonymous.

*Because it's a service provided to you.*

So, pre the Paris attacks [of November 2015], and even pre-WikiLeaks and Snowden, I am of the view – and I say it more stridently post-Paris – that I am prepared to surrender some of my privacy if I can have better security for myself, my family, my friends and society in general. I am prepared to do that. Ironically, actually, my wife was on a boat, organized by the Summit Group, which went out of Miami, and they had Snowden by Skype on the boat. There was a group of these entrepreneurs from Silicon Valley, including Google. Eric Schmidt was there. And Snowden was being interviewed by that group at precisely the same time those Paris bombs were going off. I was in Italy and I sent my wife a note asking whether Snowden was happy that his information may have caused this sort of tragedy. I have no doubt that Daesh –

*Islamic State.*

I don't think we should give them that name.

*The Caliph.*

We shouldn't. I don't think we should give them the status of calling them an Islamic State. Anyway, I just feel very strongly that this is a privacy issue. But the real issue is not privacy, it's about security. If you can get the consumer comfortable with security, I think the privacy issue sort of falls away because if I know that it's secure, I'm not so worried about privacy. But there's another thing relating to security in the sense of bomb threats and terrorists, and my view is that I want to be protected. I don't mind if the government can read my emails or look at my texts.

*Really? I do mind.*

No, no, I don't.

*That's astonishing to hear from you.*

No, I wouldn't say so – why not allow it?

*It gives me a bad feeling.*

Well, I know it's un-American.

*Yeah, it's very un-American. My students tell me the same thing: 'If you're not doing anything wrong, then why worry?'*

Why are you worried?

*Because it's your privacy. It doesn't matter if you're doing something wrong or not.*

No, I disagree with that, I really do. If I'm sitting in a restaurant, and I'm going to get sprayed by bullets from an AK-47, I'd rather be protected. When I was in Rome on that Thursday before the massacre, I was staying in the Hassler hotel. I thought to myself, 'What would I do if I were a terrorist and wanted to cause maximum mayhem? I would go into a hotel,' and this has happened in the Taj Mahal Palace and Tower in Mumbai in November of 2008, and more recently in Mali. Sitting in the hotel room, I was worried. Here I am sitting in one of the rooms on the top floor, I can't get out, and sure enough in Mali, the Westerners were in the Radisson and apparently they were in the top part of the hotel. I mean, terrible.

*The Hassler has a reputation for housing important people.*

Anyway, I would be willing to give up some privacy for greater security.

*If I could go back to the idea of profiling . . .*

Our revenue's at $20–23 billion a year. Our billings is $76 billion. We are about a third of the market. Here in the UK we're roughly the same, Germany a bit more. In India we're 50 per cent. In America we're about 26–27 per cent. On average, we are about a third of the market. And our business is affected by technology and by geography. I think we've managed to replace about 50 per cent of those revenues which fifteen years ago didn't exist. Digital didn't exist; it was just in its first stirrings. And then, the fast-growth markets were not a third of our business. They were at 10 per cent. Today, fast-growth markets are a third of our business; today digital is almost 40 per cent of our business.

In the next five, ten or fifteen years, it will shift even further. We've seen the rise of Google which is now the first or second most valuable company in the world –

*It's right after Apple.*

Yes, after Apple. The two biggest companies by market cap are both technology companies.

*That's amazing. What are your thoughts on advertising which is directed?*

Targeted, you mean?

*Targeted, yeah. It gives me a creepy feeling sometimes. I'm looking at CNN.com and I'm receiving advertisements for things I just looked up on Amazon.*

If the context is the right context . . . If, for example, let's say we're looking at your anonymized profile.

*You probably have.*

We can see that you're looking at real estate locations and houses. And we serve you an ad, which we can do.

*I know.*

We serve you an ad which lists potential real estate bargains, or removal services, or interior decoration or furnishing. It depends on the context. Perhaps we note that your car is seven years old and it's about time it was refreshed, so we serve you something that is contextually correct.

*I hear you. I should be grateful.*

Then you should be grateful. There's a guy called Ben Edelman, whom you should speak to. He's a professor at the Harvard Business School. Associate Professor Edelman has railed against Google, and he has a wonderful presentation, which I saw a couple of years ago at HBS at one of our reunions.[1] And it was classy because he's a professor at Harvard Business School. He went to his Google page live, and they served him an ad saying, 'You should be doing an MBA at the Arizona State Business School.'

*Well, it's not perfect.*

No. But it's going to get better because of predictive intelligence, which is their strategy. It's getting to a stage where it's like the seven-year car example, where we know about you, and we know or we might have some information from your refrigerator that your milk supply is coming low or you need butter. And we send you a message to say, 'Time to replenish your butter,' or –

*Yeah, Google's already doing that.*

Predictive intelligence. We go even one step further: now, this you can say is a bit creepy . . .

*I would say that, yes.*

I remember we were looking at a site with a guy who works with me here. We were looking at a site which is aimed at millennial women. This message popped up about dating sites which was totally inappropriate and contextually wrong. What a waste of time!

The answer to your question is that if the ads – and it may not be pop-up ads, it may be more sophisticated than that – are contextually right, they will win through. If I'm having a conversation on Facebook with my kids or my wife and some ad pops up which is totally contextually wrong, then that is negative publicity. If I say 'house', and I'm just talking about our house and not looking for a new house and suddenly somebody sends me an ad . . .

*For real estate.*

Yes, well, then it's wrong. We have the most sophisticated programmatic platform in the world, called Xaxis, and we link that with

AppNexus technology, and the technical reason for that is because we want to be agnostic, unlike Google with DoubleClick and Facebook with Atlas. Our view is that Facebook and Google are media companies, they're not technology companies, but they rather masquerade as media companies. They want to monetize their inventory in part (it's not the whole of their operation, they want to fly to the moon as well), but they have inventory that they wish to sell. You wouldn't give your media plan to Rupert Murdoch or Bob Iger, so why would you give it to Larry and Sergei or Sheryl and Mark? This would be my argument. Give it to us, we're agnostic.

*What do you mean by agnostic?*

If you use DoubleClick or Facebook, you don't know how the page ranking is done. You don't know how the algorithm works, because they won't tell you.

*That's their secret.*

It's like Coca-Cola's recipe. We now know what Coke's recipe is. But nobody knows how the page rankings work. We know that they alter the algorithms that determine the page rankings, but they don't tell you.

*They change them daily.*

They don't tell you, so you should know. And they drive you to their sites, they drive you, and they control the data too.

*It's a company. You don't have to use Google or Facebook.*

On search, they have a dominant share. Hence the investigation here that's going on in Europe by the EU.

*I think they'll pay a fine and just get rid of it.*

Well, it did take them more time than that. They have a problem in India, too. Yet not in China, as people think, because of Android. So they have things that they have to sort out.

*Are you competing with Google?*

No, we're not competing completely. They're a 'frenemy'. So $76 billion is our media billings. That's the money that we invest for our clients, closed through our accounts. And on that we have revenues of $23 billion, and on that we make profits of $3 billion. Of that $76 billion, this year – we just had the review with Google, they were in this room a day ago – we will invest with them $4 billion; it is our biggest media relationship. Last year it was $2.9 billion, so it's grown from $2.9 to $4 billion. Principally mobile search and mobile video. Our second biggest relationship is Murdoch – Fox, Sky, Star and News Corp – total of about $2.5 billion. Most of the big traditional legacy companies like Disney, CBS, Viacom, Comcast, are somewhere between $750 million to $1.75 billion. Facebook is the second biggest pure online company. Last year we did about $650 million. In both cases, Google and Facebook – in fact, in all those cases – we are the biggest buyer, because we have, as I've said, about a third of the market, our billings are $76 billion.

*So you're not competing, you're collaborating with them.*

Well, we're working with them and we're competing. Because, don't forget, we have a separate technology stack and programmatic platform, which is Xaxis, and we own 100 per cent of Xaxis. The technology that drives it is called AppNexus, which is an alternative to DoubleClick at Google and Facebook. And that, AppNexus, we own 16.5 per cent of. We've taken a protective stake there, and we're working with them to develop the business.

*So it's the technology.*

Yes. But we're only $30 billion on market cap. Google is $490 billion dollars. How does a $30 billion company (like a minnow) compete with that?

*I admire the fact that you're trying. Google is just huge.*

We are a motor torpedo boat in comparison to them, and sometimes motor torpedo boats are better than aircraft carriers.

*It depends on what you want to do exactly.*

For Apple now, the market cap is about $550 billion. Google is $490 billion. For Facebook it is $300 billion.

*Not bad.*

Not bad at all. So we're in tenth place. The point is that we have to be much more agile, and we can invest and we can apply technology. I talk about applying technology, so we do have some things that we've got, i.e., Turbine, which is a sort of extension of our programmatic platform, that we have developed solely ourselves. We created that technology; we actually wrote the programming and did everything.

*So you have your team of software engineers?*

Yes, we recently acquired a company called Essence which is Google's media buyer, actually, with Google's approval. We just announced the transaction a week ago. Essence has become another media agency and it has 500 people in Europe who are software programmers, basically who work with us as engineers. We are recruiting different types. This is not Don Draper in *Mad Men* any more, this is a very different business. Seventy-five per cent of our revenues is stuff that Don Draper wouldn't recognize. It's $5 billion in media, $6 billion in digital, $5 billion in data. So, you could say that 16 out of 23, which is our total revenues, is stuff he wouldn't recognize.

*Aside from making money, what do you try to achieve through advertising?*

Read the annual report. Our vision is based around talent.

*Okay.*

You said 'aside from making money'. That's not how we would describe our purpose. Basically, the big investment that we make is in people. We employ 190,000 people.

*That's amazing.*

Directly or indirectly. If you say, on average, there are three or four people in a family, that means we're responsible for well over 500,000, half a million people. One of the big controversies that you'll find out about us is that people think I'm paid a lot of money for doing what we're doing. Yet, in fact, all my wealth is tied up in this company. I only own 2 per cent of it. But 2 per cent of $30 billion is a lot of money to me. I've been doing this for thirty years and I don't sell the stock. I don't do what the Americans do: I don't have restricted stock that I sell. I keep the stock. I was interviewed by the *Daily Mirror* yesterday, and one of the questions they asked me was about my compensation. I said, 'I make no apology for the fact that we started with a company which was a million pounds in market cap, it's now £19 billion, $30 billion. We employ 190,000 people in 112 countries, affecting 500,000 people who depend one way or another on WPP and I'm proud of that. I make no apology for that whatsoever.' And everything that I earn is based on performance. And it's long-term performance, not short-term.

*What do you mean by performance? Achieving results?*

It means the performance of the company. We're rated by Harvard Business Review in the latest rankings as the fifth-best-performing company and chief executive officer in the world. The company is rated as the fifth-best performer. They rate companies not just in terms of performance (total shareholder return performance) but also on issues such as the environment, social purpose and so on.

Our purpose is very much built around talent. The mission statement is around nurturing and developing talent. Of our $23 billion, we invest about $14 billion a year one way or another in people.

*So, about half?*

Yeah, so about 60 per cent. Four hundred and fifty million dollars, we invest in chairs, desks and things. Ours is a human capital business. We really invest in talent to deliver what we do, because if we do that, we make oodles of money. But we're long term. The problem at the moment is the world is too short term.

*Could I provoke you in terms of ethical considerations? Some people think advertising is a lie, that it's deceiving people.*

No, it's not. No. You can only sell a bad product once. An argument made in Britain was that you sell people things that they don't want.

*In the United States, that's certainly true.*

No, it's not true, no. You can satisfy an emotional or psychological desire. There is nothing immoral in satisfying a psychological need. If I feel better because I wear this shirt, or this collar, or you feel better because you wear that collar, there's nothing wrong with that. There's nothing wrong because you feel better.

*I don't think anybody would debate that.*

I know, but that's what we do. We satisfy. We differentiate things for tangible reasons. One car being a better performer than another, one computer being a better performer than another. And for intangibles. What does an Apple iPhone say about you, as opposed to a BlackBerry? At the moment, I have both [laughs]. I prefer a BlackBerry actually. It's better, but for some reason . . .

*I've been told that it's better technology.*

Well, I think it's a better network, it's a more secure network. I prefer the keyboards and things but they haven't been able to make it work.

There is nothing wrong with advertising. You do get into interesting areas if you follow that argument logically: for example, does that mean that advertising should be totally unrestricted? There are some things that perhaps you shouldn't market. For another example, are you familiar with this controversy about the Lord's Prayer?[2] Have you heard about that?

*Yes. Because it may offend some. Look at the whole Christmas thing in the United States.*

This digital advertising company, which puts ads into cinemas, went to the Board of Censors because the Board of Censors has to pass any film, including ads. The Censors passed it, but the advertising sales company said no to this ad. It was done by the Church of England and it does contradict the advertising sales company protocol. Their protocol says that they will have no political or religious advertising.

*Well, there you have it.*

Yeah, except they did allow ads for the Scottish independence party, during the Scottish referendum.

*Which is political.*

Which is political.

Personally, I don't mind somebody reciting the Lord's Prayer. I think I told the *Daily Mirror*,[3] 'I'm Jewish, but I used to recite the

Lord's Prayer – Our Father who art in heaven – every day in school.' So I've got no problem.

*You went to a Christian school?*

I went to a school called Haberdashers', which had a lot of Jewish boys in it. In fact, I used to take Jewish prayers. In the primary school we used to say the Lord's Prayer except I didn't use the word Jesus; if it had Jesus in it, I excluded it.

*Well, the Lord's Prayer does not have the word Jesus in it.*

It doesn't. That was okay as far as I was concerned [laughs]. But in the senior school, I was a prefect and I took Jewish prayers. We used to do the Shema and the Amidah and this and that in Jewish prayers, and then we used to get wheeled into assembly for the announcements.

There are obviously some limits in terms of what you advertise, but basically David Ogilvy said, 'The consumer is not a moron; she's your wife.' Politically incorrect. You should say now, 'The consumer is not a moron; it's your partner' [laughs].

*Why do they always sell cars with very pretty young women next to them?*

Because that satisfies probably an emotional desire.

*It obviously does.*

But they don't always do that. In fact, for our Ford advertisement – Ford is our largest client – we don't use that concept. We stress functionality. We go further.

*Well, it depends on what audience you're targeting, I guess.*

No, but even when you go to the Detroit Auto Show, you will find attractive women, yes, in skimpy outfits. But do I think there's anything wrong in that? Well, if you start to move to restrict that, you're going into some very dangerous territory. If, for example, you said, 'You cannot have skimpily clad women at motor shows.' Well, how far would you go? What constitutes a woman who is not skimpily dressed – or a man?

*Or at Formula One races.*

I'm on the board of Formula One, right. If you remember, at some Grand Prix they've had men as the assistants (instead of young women). You get into dangerous territory with censorship. That doesn't mean I think things should be completely unrestricted. After all, here I am arguing that the government should be able to have access to more information to increase security.

*Does WPP have an ethical standard? Does it have a set of principles?*

Fully laid out. There are three books there; one is the annual report, in which we do a statement about the industry. The other two books are really interesting, because the first book looks at what we do in terms of purpose with our clients. And then the other looks at what we do ourselves. So we do three books. It's online as well, and we've won all the awards. Here, you'll see in both these the pro bono work that we do. It shows what our agencies around the world accomplish: work that we do at no cost. We do wonderful stuff. We do wonderful things, even as I say so myself. It's superb. This is what our people do on behalf of the needy. Sometimes it's funded by clients. Take, for example, what we did in Turkey: we sponsored Women's Day to prevent women being abused and beaten up. We created an app with which, if any woman was worried, she could just notify the police. And it's working, because the guys know it. Absolutely.

*That's marvellous. I didn't know WPP was involved.*

We're not dummies. This is what we do with our clients, and ourselves, on sustainability. It's all contained here [points to the books]. We've taken the lead in it.

*Obviously.*

We're the only company in our industry that does it.

*Here's a question that's probably not in one of the books. In your opinion, does the Catholic Church need help with its advertising?*

Well, you keep on using the word advertising rather than communication. Does the Vatican need help with communication? Yes. Well, mind you, you have now a Pope who is a rock star. He's a Modi or a Macri.

*Certainly in the United States.*

No, he's a rock star everywhere. So he's a really interesting individual. I do think leadership is absolutely critical. The difference is what the new Pope has made – a dramatic difference – in a very short period of time. He has good people advising him.

*True. What would Sir Martin do if the Vatican called and said, 'We need to project an image. We need to have people perceive the Catholic Church in a certain way'?*

You can't answer that question unless you do some in-depth work on it. It is certainly fascinating. We're working for a number of countries at the moment. And we work with cities. And we work with mayors.

*Countries in the sense of how they're perceived?*

Yes. Countries.

*Like North Korea, which is probably not perceived in a very good light?*

Yes, but for North Korea we would have reservations taking on the assignment. We're ready, like we are in Cuba, where we just signed an agreement with a Cuban agency.

*So, that's now open.*

Yes. We have a man in Havana, like we had in Myanmar and Vietnam before that. Coming back to the Vatican, I think there are some things that could be done to modernize and contemporize the Catholic Church. Pope Francis has actually done that. I think it is remarkable what he's done already, with or without professional advice but with good people around him; quite extraordinary.

*He's very charismatic.*

Oh yeah. He is. He is a rock star. He might not like to be called that, but he is a rock star. And Argentinian. I like the fact that he gets about in a Cinquecento, rather than the Popemobile. I think the whole idea, the whole task, is really interesting. How do you contemporize an institution like the Vatican which has gone through a lot of problems and is often criticized? Concerning the structural or the organizational changes that were made – I don't know what the real story is. You could say the Vatican is a legacy business, right? And you have disrupters –

*I'm not sure what a legacy business is.*

A legacy business is a traditional business.

*We've been doing this for 2,000 years. I think the only people better at this than us are the Jews, because they've been doing it for longer.*

But you have disrupters. You can say Daesh is a disrupter in a way. You have on one end zero-based budgeters. I don't know who the zero-based budgeters in religious terms would be, but in commercial terms it's these companies like Anheuser-Busch, that stress cost. And then you have the short-term investors like Nelson Peltz, Bill Ackman and Dan Loeb bearing down on you, telling you to look at the short term. That's why I complain about people being too focused on the short term. But there are disintermediaters, disrupters to traditional religions – Judaism, Catholicism, Islam. Daesh is basically saying we have to be fundamentalists, that a modern interpretation of the Muslim religion doesn't work.

*They're convinced.*

There's tremendous disruption. So, how does an organization as well resourced, as well respected as the Catholic Church change? To what extent does it change? How does it change? How do you implement the change successfully in a world that is changing?

*What do you think of the Catholic Church's response to the sex abuse scandal?*

When there is a case, the media story is top of the agenda, but you can't say it's the media's fault. You never win by not communicating. If a journalist makes the call, you never win by your PA saying, 'I'm sorry, he's travelling,' or, 'He's in a meeting.' If there's one thing that drives me crazy when I talk to one of our people it's to be told: 'He's in a meeting.' I say, 'I don't care what the bloody hell he's doing, I want him on the phone.' I don't care.

*Well, if you're Sir Martin Sorrell . . .*

No, because I wouldn't call unless it was something I needed an answer to. Similarly, if somebody calls me, doesn't matter who it is,

I try to respond. Or somebody sends me an email. One of the things I'm notorious for – if that's the right word – is responding very quickly to emails.

*Yes, I was completely taken aback when I got your quick answers to my emails.*

If somebody writes to you and takes the time to write to you, you should respond. But you have to engage. You cannot bury your head like the ostrich in the sand. You have to engage.

*Did the Church not have a good response to this crisis?*

They didn't respond. They didn't engage. You don't necessarily have to hire professional people but you've got to have people who look at the problem and give advice.

*Why do you think that the higher-ups in the Catholic Church don't pay very much attention to this kind of thing?*

I think they're worried about engaging. I think they're frightened of engaging.

*Isn't that worse?*

Yes. But corporate executives are the same. I think they should engage. Not necessarily because I'm CEO of WPP, but because I'm an individual and I believe that saying what I think is the right thing. I shouldn't be muted or silenced by virtue of the fact I'm speaking in a personal capacity.

I think institutions are inherently bureaucracies. They're slow-moving. They don't have decisive leadership. You get these matrices that prevent them from making decisions. Everybody has an opinion: 'We shouldn't do this, because of this.' So by the time you're finished it's what you call in Yiddish 'a matzoh pudding'. It's a mess.

*We call that Vaticanese. Vaticanese is a way of saying things that appeases everyone and offends no one.*

So it's very difficult to make a decision. That's why you need strong leadership. The fundamental point is this: you never win, in my view, by not communicating. For example, we didn't go to Sochi for the opening of the Olympic Games, Obama didn't go, Cameron didn't go, Merkel didn't go, Hollande didn't go, none of them went. Xi Jinping went, Abe went, Erdoğan went. When we didn't go, we upset Putin to a far greater degree than was valuable. It was his personal project. He was supposed to have spent $1.5 billion on the opening ceremony. It was a great opening ceremony. Closing ceremony was also excellent. He included a lot of people that you would never think would be in there, but we insulted him and this was the personal point. We pay for these people to be diplomats and to be politicians. We should have been talking. I'm not saying we would've avoided the Ukrainian crisis, Crimea and all that – but we would've had a better relationship. You don't win by humiliating people publicly. You win by engaging. You win with China by private diplomacy, not by haranguing them about Tibet or the Dalai Lama – or whatever – in the press. The Chinese do listen and they do learn. I think it's their greatest virtue. It's our third-largest business, we have 16,000 people in China. We've been in China since we've started and we have the most powerful advertising and marketing business in China. It's much bigger than anyone else, domestic or foreign.

*Really? Even for the Chinese themselves?*

Yes, $1.7 billion.

*That's amazing.*

We've done a great job so far. In India we're 50 per cent of the market. We understand the BRICs, I think. In Russia we have the leading business, and in Brazil too. But the point I'm making is this:

you win by engaging, and believe me, they listen. That has been my experience.

*Are you convinced that WPP is a vehicle of good values for the world?*

Yes, absolutely. Are there things we have done that I would change? Perhaps, but by and large, I think we certainly provide a good living in many ways, shapes and forms for the 500,000 people who depend on us. I think we do good things politically, socially, economically. There are some things that sometimes are difficult for us to do, but we try to embody the positive outcomes. We try and do things, whether it's the pro bono work, COP21 and climate change or the Secretary General's seventeen sustainable development goals. We're doing a lot. I mean if you get a chance to read about that stuff . . .

*I will. I will read that.*

It is impressive.

*I interviewed the director of the national news in Italy who is responsible for placing news items, for choosing which ones to run and in what order (which as you know is the key to making the news). I said, 'Are you aware of the power that you have?' And he said, 'Yes, we can destroy people.' Does WPP have that same power?*

No, I don't think so. We have the power to build: we have the power to help companies build their reputations. So, by definition, I suppose you could say we have the ability to do the opposite.

*One would assume.*

The question is whether we do it or not. Do we do negative advertising and have we done negative advertising? Yes, historically. A lot of people think (for just practical reasons) that negative advertising doesn't work. If you mention your competitor, you give them free publicity. It is not very effective. But do we have the power to build things up? Yes, we do. We do have that power. Whether we have the ability to destroy, I'm not so sure. And whether we would do it or not, I'm not so sure. But we certainly do have the ability to amplify. I think we can only amplify good things, only amplify things that resonate. It comes back to what I just said before. You can only sell a bad product once. But the consumer – David Ogilvy's right, the consumer is not a moron.

*Cigarettes?*

Yes, we do do cigarette advertising.

*It's not a very good product but it sure sells.*

Yes, but our view is that the consumer decides – the consumer is educated, and the consumer can decide. The problem with these restrictions is, once you start to place them in the industry, where do you draw the line? What about fatty foods? Where does it begin and end? And it's not just about fatty foods; what about exercise as well. For example, are people who produce electronic games unethical because consumers are sitting like couch potatoes and drinking sugar drinks and eating hamburgers whilst they're playing the video games?

*No, the point is well taken. It's not their fault, it's the fault of the individual.*

Society is changing. Now, if I look at the habits of millennials, people born from 1985 onwards (so up to thirty years old), or centennials (who are not 100-year-old people, they're people born from 1997 or up to eighteen years old), their habits are very different. Centennials use Snapchat, millennials use Facebook. So why do they use Snapchat? Because they don't want their parents seeing what they are doing. Snapchat is like mission impossible: it goes away.

*Well, supposedly. One of my final questions is always what do you see coming within the next five to ten years in your sector?*

I see more fast-growth markets, despite the fact that the BRICs and Next 11 are now a vice rather than a virtue because they're growing more slowly. The future of this company depends on our penetration in those countries. Long term, these countries are going to succeed. The next billion consumers will come from Asia. The Chinese have modified the one-baby policy, but I don't think that's going to imply a massive increase. Within twenty-five years, India will be the most populous country on the planet. It will have the third-biggest GDP. China will have the biggest GDP, but will be the second-biggest country by population on the planet. USA probably will be the fourth-biggest, because Indonesia will have overtaken the US. The largest GDP per capita will still be in the US. But the biggest will be China by GDP, second will be USA, third will be India. The future of WPP depends on that, number one. The future is digital so 37 per cent of our digital business should be increased. Data will be even more. Those anonymous profiles will become even more important and targeting them will become important. And last, in working with our clients, integrating our offer into what we call horizontality, which is probably not English (although it is in the annual report). Horizontality is getting our people to work together. Because we're very vertically organized by brand. Like most brands.

## 7.
# *Publicity*
## (*Maurice Lévy*)

Centuries ago, when someone encountered a person of refined taste and elegant clothing, knowledgeable about current events, able to speak multiple languages, evincing a vast cultural understanding, acutely aware of his own place in history, that person would be called a 'Renaissance man'. This is how I would describe Maurice Lévy. Since 1987 Maurice has been the Chief Executive Officer of Publicis Groupe, the third-largest publicity group in the world. It has a market cap of €13.64 billion as of 2016.

Maurice invited me to speak with him at the Publicis headquarters in Paris, close to the Arc de Triomphe. We ventured into many themes that afternoon, and I believe his insights on these important topics serve to illustrate his status as a modern-day Renaissance man, who is busy helping people to communicate in our world.

* * *

*Could you briefly tell us, who is Maurice Lévy and how did you get involved in advertising?*

I started my career as an engineer, a computer engineer.

*Oh, so you have experienced this digital revolution first-hand?*

I was programming algorithms when I chose to go into advertising, because it was more fun [laughs].

*You did not know that the future was going to be digital at the time, did you?*

No, no. However, I realized that the future would clearly centre on computers. I started my career by running an IT department in an advertising agency, instead of going to other places with much larger computer rooms and higher offers, yet where the environment looked a bit bureaucratic and much less fun. To put it another way, joining the advertising world was not by design but rather by accident. The day before I was about to choose to work for the Commissariat à l'énergie atomique (Commission for Atomic Energy), I received an offer from an advertising company and decided to work with them. I was happy to work with people of my own age wearing jeans and with long hair. This was in 1966. I worked in that agency for five years and then I joined Publicis Groupe in 1971. Here, I stayed and eventually rose through the ranks to the top. It's that simple: I love advertising. Working with formidable teams that come from all horizons keeps you on your toes, it nourishes passion and I must say that I never ever got bored.

*Your area is publicity, advertising or communication. Sometimes your industry gets a bad reputation because people think that advertising makes us buy things we don't need. Perhaps you can give me your thoughts on that.*

Advertising is an integral part of a market economy, a key element. The role of advertising is to develop mass production/consumption by making products accessible to the greatest possible number of people. Without advertising it would have been almost impossible to have products such as home appliances, cars, washing powders, etc. – all the attributes of progress and modern society – accessible for the vast majority of people. Thanks to mass consumption, product democratization leads to affordable prices. Advertising also makes the competition fiercer between manufacturers, for the benefit of consumers. When you look for instance at some technological products, from TVs to smartphones, you have an incredible concentration of technology which would have remained the privilege of the happy few without advertising, its mass distribution and the associated brand competition.

Yet I agree that people sometimes criticize such a society, either because they are specifically against advertising – what we call in French *publiphobie* – or because they are simply against this kind of consumerist society. I fully respect these points of view and I also concur that some exaggerations exist. We have to think the way consumers are thinking, i.e., look at any product through two dimensions, at least: on the one hand, the objective measurement of the service we can expect from a product; on the other hand, the fact that owning a brand or product actually gratifies people. We should never discount the merit of such a psychological gratification, as that is part of what we owe to our clients: beyond the intrinsic qualities of a product, they need to differentiate from the competition. Fundamentally, this is what advertising does: create differences, which lead to preference.

To deny advertising amounts to denying the idea of progress, which is a rising tide that lifts all boats. To a certain extent, being judgemental towards people and their ability to take things into consideration is also an insult to their intelligence, their freedom and their desires. I see no reason for only fulfilling the desires of the happy few. And if some middle-class people want a luxury product from time to time, why should we prevent them from owning it if they can afford it?

Moreover, advertising is instrumental in making some key services or products available for free. Think about news or free media such as TV and radio – as a sustained revenue stream, advertising supports democracy and freedom of opinion. Without advertising, the Googles and the Facebooks of the world could not have become what they are today, their reach (above one billion daily users) would have been much smaller. The whole internet model is actually built on free services for advertising.

*Those services don't charge us money to use them, but those companies end up with our data.*

That is something that comes afterwards. Without advertising, they would not have received your data in the first place, because you would not have used their service: you would have had to pay for it.

*Yes, that is right.*

It is the magnitude of the number of users that makes our data valuable. If they had to ask users to pay in order to use Google or Facebook, they wouldn't have reached such a scale. As a result, monetization of data would have been purely theoretical. With data, we have to work on a new protocol in order to protect privacy and make sure that data is properly used.

I would like to come back to the advertising/media couple. Media guard their independence and the independence of content jealously – and it is paramount to protect this. Yet, strangely, advertising allows this better than any other owner (think about the state-owned press). Again, freedom of the press and democracy are linked to advertising. Advertising fosters the plurality of opinions and at the same time offers an entertaining break. Like any other industry, we can certainly find ways to criticize advertising but the role of advertising in modern times is indisputable, beyond some criticisms that could be made. Our society would simply not have been the same without advertising: no mass production, less progress and the privileged services reserved to a small elite group instead of being shared by the vast population of the world – and probably press propaganda instead of plurality of sources.

Working in an ad agency is truly rewarding and fascinating because you can work in almost all sectors of society: food, beverages, packaged goods, cars, beauty, clothes, fashion, media . . . You have access to politics; you witness first-hand the importance of sociology and psychology; you understand the big trends of all industries; the relationship between the manufacturers and retail; you are in the middle of all the fights and competition that exist between brands. Furthermore, you deal with highly sensitive and yet such captivating feelings: human emotions. We are constantly working with feelings and emotions. We define our company as the alchemy of creativity and technology, as the combination of IQ plus EQ plus TQ plus BQ, all these factored by CQ. IQ stands for intellectual quotient, EQ emotional quotient and TQ technology quotient. BQ because this world is moving very fast and you have to be bloody quick!

*I did not know what BQ meant [laughs].*

And CQ because all of this is factored by the creative quotient. This is what we do at Publicis Groupe and today we are on the edge of a transformation. In advertising, we are a group of craftsmen, what we do is all about craft; it's about art; it's about spoken words; it's about ideas; it's about emotions; it's about creative strategies. All the campaigns that we create are built and tailor-made for a client, tailor-made for a product. Each and every campaign is a prototype that you can't simply take off the shelf. One might say, 'I will take this idea which I have not sold to a client and will sell it to someone else,' but this doesn't work. We have to think specifically for each client. And what I love above all is the incredible combination of a strict rationale and fantastic intuitions.

In the new world, on top of everything that we are traditionally doing for a client, we need to take into account new technological and digital tools – which leads to the use of technology platforms. Our clients face lots of challenges and most of them think about transforming their business models. If you add to the mix what data and artificial intelligence will allow going forward, you can imagine what the future could be. We therefore need such technology platforms, something that is really embedded into IT, where we help our clients transform themselves. Also, data plays a key role. To be honest, we have always had data involved in what we do, yet never as much as today. At the end of the 1970s, I authored a chapter in a book on data and how to deal with it.

*That was over forty years ago.*

So, you see it's something that has always fascinated me. Look at people who were doing direct marketing: they were already working on data. They were already using client lists and they were trying to implement a new approach regarding how to best target people, to make a profile as fine tuned as possible. They did not have the incredible mass of data (maybe too much) that we enjoy today.

There are lots of pieces of information shared on Facebook and other social networks. These pieces are important to individuals – part of their lives, actually – but not really important for society at large. When I think that all this data is stored in the cloud, I believe we should rather have redeemable data that is deleted in due course. And I tend to think that most people would agree: think about Snapchat that is a self-deleting messaging system.

There is another interesting cloud to discuss: the one made of the words you use when communicating, searching, blogging, etc. The combination of words builds different profiles from the ones we can assemble using socio-demographic and economic data. Such profiles are far more interesting because we can define individuals not only by some hard numbers or hard facts (such as your age, whether you are single or married – in your case single clearly) . . .

*At least up until now, yes.*

We will see what happens [laughs]. We can build much more precise profiles for people.

*Would you be allowed to tell me who are some of your clients?*

Yes, our largest and number-one client is Procter & Gamble.

*Oh, okay, a small company [laughs].*

A few other small companies such as Nestlé, Unilever, Toyota, General Motors and also Philip Morris, amongst many others.

*They are the number-one manufacturer of cigarettes, which is not a very good product for people's health.*

Health is a paramount issue that we have to take seriously and cautiously. This is one of the reasons why advertising has been banned and communications on tobacco are strictly regulated. As long as the product is legally authorized, I consider that we must accept to work for them. Obviously, people are free to choose to smoke or not, and talents within our organization are free to work for such a client or not.

*Do you have mobile telephone companies as clients?*

Yes we do. We work for many companies around the world, notably Orange, Verizon, Telefónica, T-Mobile, etc. And we also work for mobile phone manufacturers such as Samsung, which is one of our largest clients.

*Samsung is huge.*

Indeed, and we are extremely proud to be working for such a large company, such a great brand that makes us grow on a regular basis. Publicis Groupe is the third-largest communications and advertising company in the world and the most advanced in technology, by far. Such an achievement has been underpinned by our clients' trust – including the trust of very large clients such as Samsung.

These clients continue to trust us with their future. We recently acquired Sapient, a start-up that was born in Boston and grew up with the internet. They are one of the cutting-edge operations in technology, data and consulting.

Our world is changing dramatically. We have learnt from the new players and observed the evolution of clients' needs. We decided that we were at the end of advertising as we knew it – a book title I borrowed from a friend of mine, Sergio Zyman, for the keynote I gave during an industry gathering in Miami. The end, because it is about our relevance in this brave new world – about our own transformation. Paradoxically, the ad world is one of the most conservative sectors you can imagine. People are attached to the ways they have been working, to the silos, to the systems, and so on. At Publicis Groupe, we decided that the best way to help our clients on their own transformation path was to transform ourselves. Not an easy task.

*I would like to provoke you on two topics. One, are you allowed to tell us about your relationship with Omnicom and John D. Wren? There were rumours of a merger between your two companies.*

They are not rumours. We actually announced that we had signed an agreement to merge. In a world where scale matters, we felt that a big merger would help. We came to the conclusion that the only possible merger was a merger of equals. Unfortunately, this did not work and we decided to call the merger off. I remain a good competitor of John Wren. We have always been cordial competitors and we remain so. The merger was terminated amicably and the expression I used was that it was better not to go to church rather than go to the judge for a divorce.

*May that happen in the future?*

No, no. The two companies went in different directions and the best way for us to compete has been to each continue in our own way. This merger was contemplated before acquiring Sapient, as in those days Sapient was not open to a transaction. Since we acquired Sapient, our profile has changed dramatically: 50 per cent of our business comes from digital technologies and we are not ready to dilute it.

*Fifty per cent?*

Yes.

*Amazing.*

Today, it is no longer worth thinking about the merger, and we are building a future for us and we are transforming our business in a significant way.

*Here is the second provocation: do you consider Google a friend or an enemy?*

Google as well as Facebook and many others are our partners: we work with them to best serve our clients. They are partners and at the same time they are competitors in some small areas of our business. They will hardly be competitors at the core of our business, i.e., creativity, as creativity is a specific craft that requires talented individuals and cannot scale. Whereas the models of the Googles and the Facebooks do scale by design. They develop tools to be replicated and used endlessly. We develop prototypes and need headcount. The day they move to headcount, their margins will shrink dramatically – which they will certainly not welcome happily since their valuation multiples would take a hit. For this reason I believe the competition will be rather limited. However, there are new competitors who are coming. Just look at the press and you will see that IBM is purchasing digital agencies, Accenture is doing the same, Deloitte as well. People who were not even on the same planet as ours are now landing on our planet. That is also one of the reasons why I have decided to add consulting and technology to our services, so that Publicis Groupe can offer an end-to-end service to our clients, from A to Z: from building the marketing platform to the transaction directly on e-commerce, through media and creative.

*Okay. I would like to explore now a sensitive theme which is close to my heart: ethical or moral considerations. Is Publicis only about making money or do you feel that you help people in some way?*

You should have been at the debate that I had in Davos with the Booth School of Business from the University of Chicago (and you know, the Chicago school of economics is all about maximizing shareholder value, as expressed by the Nobel Prize economist Milton Friedman). We had a heated discussion at Davos where I defended the view that a company today cannot be only about shareholder value, or only about profit. We have a responsibility to our employees, to our clients, to our suppliers – we have to respect and treat our suppliers extremely well. We see that sometimes we are squeezing the suppliers too much. We have a responsibility to the city. When you put all these together, there is a notion that my friend Professor Schwab, the founder of the World Economic Forum, coined: the 'stakeholder value', which is no longer simply shareholder value. We have to be responsible to all of our stakeholders, not only to our shareholders. Our employees have a stake in the company that is at least as big as the shareholders' stake. They contribute to building the company; they contribute their passion, their sweat, their talent, in order to create something big. There are many other aspects that we need to take into account, notably Corporate Social Responsibility (CSR).

*And how many employees do you have now?*

Roughly 80,000, all over the world. We are in about 100 countries. We don't always have shares in the countries where we operate. Sometimes we use partners to represent our interests. The decision depends on many factors.

*Here is another ethical consideration. You mentioned before how important your business is for all of society, not just for some products but that it is really transversal. Are you aware of the power that you have?*

I think the power attributed to advertising is a fantasy. Some people believe we can mesmerize the eyeballs, change the minds and sell ideas – using some kind of invisible power. The reality is fairly simple: people accept and understand the role of advertising when it comes to products and commercial propositions. They are more sceptical when it comes to ideology and politics. We must be extremely professional in the use of our craft, capabilities and technologies. That is why Publicis Groupe is not willing to work for political parties or candidates. I believe that it is not only a misuse of our capabilities but it is also a risk to divide the agency itself – as, hopefully, we have a fair representation of ideas within our organization.

When you carefully examine the details, the reality is very complex. First of all, consumers are empowered. We must respect their intelligence. They understand what we do. They know how to decipher a campaign and they understand the underlying technology used in marketing. On top of that, they have the power to react. The so-called power of advertising is best used when communication is honest: it is about trust, true values and good deals for the consumers. Any other approach would damage the brand and be against the interest of the client – and it may destroy value. We therefore have to be extremely, extremely cautious. This doesn't mean that we shouldn't dare to use bold creative ideas but we must be dead honest.

*Okay. It's obvious that as a company you are able to build up a client. You are able to support and have a client expand and be successful. Do you have the same possibility of destroying a client, a product or a brand?*

If we create the wrong campaign, although it would be unintended, we could ruin a product. However, if you are referring to the intentional destruction of a product or a brand of one of our client's competitors: this is something that is prohibited by law and by the code of ethics of the profession. You just cannot do that. Unintended mistakes or unintended negative effects have always existed, but deliberately destroying a product or a brand is not the goal of advertising as we see it. The goal of advertising is to create value through the brands you are building and defending.

*Look at the classic example of Pepsi-Cola and Coca-Cola.*

Yes, but don't tell Pepsi-Cola that their product is the same as Coca-Cola. They will tell you that there are many differences between the two products. Since you mentioned power, if there is a real power in advertising, it is about creating difference through creativity.

*I recall in the 1970s there were negative commercials about different types of soap, such as Ajax, that tried to bring down the competition with negative advertising.*

Yes, but that is comparative advertising: here you are not destroying but you are trying to state a comparative advantage compared to another brand. You have to be absolutely certain that what you are saying is absolutely true. If you state something that is not correct, the competitive brand will sue you and the damage which will be done to your brand will far exceed the advantage that you wanted to gain.

*So, it's not done very often?*

Comparative advertising exists and what we call *comparable* advertising exists. Most of the time it is with unknown brands, not named brands; but with named brands, you absolutely need to have a characteristic trait, which is meaningful to the client, and that you can claim gives you a clear advantage. Take, for example, disposable nappies: you can claim that this nappy absorbs far better and that the baby that wears it does not get wet. If you can prove it, you have a clear competitive advantage and you have to go for it. For instance, regarding detergents, you have the whiter-than-white pitch; obviously, if you cannot prove it, it will not work.

*Talk to me about the values that you want your company to transmit. You already mentioned several: you want your employees to be happy working for you, you want to be honest and fair with your suppliers. So, money is not the only priority. Are there other values that you want to see transmitted?*

While we believe that success and profit are important – it is business, and if you don't generate profit as a competitive and state-of-the-art entrepreneur, you cannot invest, and progressively you go into a negative spiral that ultimately gets you out of business – we also believe in strong values.

One of our founding beliefs is to believe in mankind, as we say in French: *nous croyons en l'Homme*, we believe in people and talents – and this translates badly in English but the French expression is very strong. Hence our mantra, 'Viva la difference', and this phrase actually consists of three languages: 'Viva' is Spanish, 'la' is French and 'difference' is English. It shows how important differences are to us: differences of race, religion, behaviour, sexual orientation – you name it. We respect our people for who they are, their culture and the countries where we operate; we have a deep respect for the consumers and the brands.

Respect is something that is extremely important and embedded in our culture. Obviously, we believe in the power of creativity and we believe that a big idea can actually move the needle. We also believe in honesty, integrity and good citizenship. We think advertising can help make the world a better place – a little. How do we do that? Through a lot of *pro bono* operations, for instance against cancer and other diseases, helping research. We support Amnesty International and various other causes.

To respond to your original question, we believe that we should not be involved in preferring one religion over another. We should not try to use our knowledge to convince people to do something religiously or philosophically. At the same time, we respect differences and we believe that it is very important that people feel good about their beliefs and their convictions. I realize that this is a very thin line. Personally, I am a member of the support team of the Collège des Bernardins. I don't know if you are familiar with this Collège?

*I have heard of it but I have never seen it.*

You have to visit the Collège while you are in Paris. It is from the thirteenth century and had been abandoned and even became a fire station at one point. It was rebuilt by Cardinal Lustiger and some friends. It is simply fantastic. You should take an hour and visit the Collège. And with your priestly garb, they will let you in no problem.

I am also involved with my own religion and I support social Jewish organizations. And the most important aspect, which I hope to see blossom one day: I am a great supporter of the Israeli–Palestinian peace talks and a supporter of the Peres Center for Peace. As you can see, I do love differences.

When we do advertising, we try never to hurt people and to bring some psychological reward – the pleasure of beauty and of an entertaining work.

*If you were aware that one of your campaigns was deceiving people for the benefit of a client, what would you do?*

I'm just thinking if this has ever happened: 'deceiving' is a strong word, it means almost 'betraying', much stronger than 'disappointing'. Actually, I have stopped a campaign, as things can change and make a campaign ill-timed or unsuitable. During the recent financial crisis, I recommended to many of our financial clients to stop the campaign because times had changed and it was inaccurate to say certain things to people the way they were said. There have been some other examples as this is something that we do when we feel that our communication is inaccurate.

*One example I was thinking of (that is not necessarily related to advertising) involves Volkswagen, when they specifically deceived us about emissions from their engines. I don't think they advertised that, but I think there was deception, inasmuch as they were manipulating the numbers, which is against the law, but it is also unethical.*

I agree with you. The only thing they have done in terms of advertising was to take out the words 'Das Auto' – 'The Car'. I don't believe it is sufficient, and if we were running the Volkswagen campaign we would have proposed to react very differently. I think that when you have been advertising so strongly for years about the reliability of your product, and now that consumers have the feeling that you maliciously betrayed them, you cannot continue on as if nothing happened. I think their apology was not congruent with the damage that was inflicted.

*I think you're right. I think it's very expensive for them so they are trying to go as slowly as possible.*

There are many aspects to the issue. You have the ethical point of view, and the first movement such as apologizing or recalling the product. Then you have the lawyers who say, 'Oh, if you do something like this, you will admit that you are guilty and you will incur a lot of class-action suits. So, you better stay quiet; don't do anything.' There are always trade-offs between being extremely cautious, based on a legal analysis, and being in sympathy and empathizing with the consumers. However, if you want to have this empathetic relationship with consumers, you absolutely need to look at things differently. Being too cautious over the legal aspects may not be advantageous from a commercial point of view but we live in a society that is governed by lawsuit-ers.

*In the United States the atmosphere is very litigious, but I'm not sure what it's like in France.*

In France, the environment is becoming very similar to the US. Litigation is less intense than in the United States, but we are catching up, unfortunately. I think we are making a wrong decision and we are polluting society by becoming far too legalistic and litigious. All this leads to a society that is not governed by principles but by rules-savvy lawsuit-ers. I obviously appreciate a country that is governed by the rule of law. Yet, we tend to replace principles by rules, meaning that you can flout the principles as long as you abide by the rules. This is dangerous as it leads to bypassing the rules smartly, assuming you will not get caught or you can defend yourself. Look at people who have cheated for years without getting caught; for example, Bernard Madoff, or those involved in the Enron case: they were difficult to catch because technically they were superficially respecting the rules, or at least people thought that they were respecting the rules, and the authorities could not detect any wrongdoing. Therefore, sometimes it is much better to have stronger principles and to rely on principles rather than on rules, systems and processes.

*Thank you, Maurice. I wish you well in your many pursuits. It is very clear that you do love 'differences'.*

## 8.
# *Cyber Security*
## (*Dave Aitel*)

Getting someone to talk with me about cyber security was not easy. Several experts that I have met these past months refused to be interviewed or go on record. This is why I was so happy that a former NSA software developer who now runs his own security company was willing to converse with me about this sensitive topic.

Almost everyone I speak with about digital technology stresses the importance of cyber security. People and businesses (as well as entire nation states) need to feel that their sensitive information is safe and secure. Dave Aitel was convinced many years ago that in the new digital era, security was going to emerge as the most crucial area of concern, which led to the founding of his company, Immunity, a supplier of security products and devices on a global scale. Operating out of Miami, Florida, but with consultants and engineers around the world, he guarantees safety for his clients.

* * *

*I usually start these interviews by asking the person to tell us a little bit about him- or herself, so who is Dave Aitel?*

I grew up in Fairfax, Virginia, which is quite close to Washington DC. I went to a technical university called Rensselaer Polytechnic Institute, and my university education was actually paid for by the National Security Agency of the United States.

*Wow. Yes, they are in the news often.*

Yes, they do appear in the news; and I worked at the NSA afterwards for several years and then I ended up working at a company called At Stake, which was a security consulting company. That was in New York and shortly after 9/11 I left to start my own company called Immunity, which is where I currently reside.

*And what is Immunity?*

Immunity provides products and services, and in terms of products we have one called 'Canvas' which is a penetration testing tool: essentially, it hacks into computers. In the services area, we also do a 'for hire' hacking into your computer for you and then we write you a really long report about it, and this is called penetration testing. There is a really good television series on right now called *Mr Robot* and that is what we do.

*Okay.* Mr Robot – *is that on HBO?*

I believe it is on the USA network. The first one is free on YouTube and it is well worth the watch because it is the only show about a hacker that's ever got it right.

*Really?*

It is amazing. It surprised everybody, because nobody believed they could get it right and yet they did.

*Wow. I have so many questions that it is difficult for me to choose so I'll just go ahead and shoot them out. You mentioned the word 'hacker': what do you mean by that?*

You know it's funny because I think of hacking as a discipline; it is similar to asking yourself what does it mean to be a Kung Fu master? It is a discipline that you have applied to your life. I find that hackers in a way have a very particular mindset and a particular discipline which carries over to other things that they do in their lives, and which is sort of shaped by their long interaction with computers at a level which most people never really get to, and which is probably unhealthy.

*Absolutely, absolutely.*

What is a hacker is a very good question. It's a very complex question but the reality is that there is a large community of people who have gathered together and who have all devoted themselves to a particular set of problems; and those problems have shaped them, even though they were the ones shaping the problem as well. At that point, the abyss has looked at men and has changed who they are. You can perceive certain characteristics of hackers in all parts of the world that will be the same, and while some of it is predictable, other parts you would never imagine to be true, but they are. There is a shared ethical and moral set of values; there is a shared almost historic culture of things that they find funny, and other people would not find funny. There are comics that they find funny that other people do not find funny. In the larger sense there is the whole computer science field, and in the smaller sense there is the particular discipline of hacking: the transcending of boundaries and becoming an iconoclast and applying that discipline of the iconoclast to your everyday life.

*That is pretty profound. I wasn't expecting quite such an articulate answer, but thank you! That certainly is fascinating. Hacking is really a mindset; it's not just an operation, is it?*

It is almost like a daily behaviour, like people who are into yoga – they do yoga every day. For people who are hackers, you have to remember that you're generally hacking something every day. Just as yoga controls your breathing, hacking can control your mentality, it changes your personality and who you are. I'll give you an example about me. I'm sure you've read the book *Guns, Germs and Steel* (by Jared Diamond, 1999).

*I have not read that, no.*

It's a good book in the sense that it won a Pulitzer Prize for providing a historical analysis of the factors that cause some societies to succeed and others to fail; and the primary reason why they succeed or fail is access to food, and the quality of food. For example, once I had read that, I then had to know where all food came from. And then you go to a Chinese restaurant, for example, and they are serving Kung Pao chicken, and you start wondering and say, 'I don't think Kung Pao chicken comes from China.' It turns out that most of the food we recognize as Chinese comes from San Francisco. Take fortune cookies, for example: they come from San Francisco. Of course, a lot of the food we think of as Italian actually comes from China. Once you go down this path, hackers learn that they must analyse these realities and at the same time it annoys them. For example, it annoys them when people use red peppers in Chinese food, because red peppers can only come from Mexico: it's that level of annoyance at the way that the world is misrepresented which is a particularly geeky thing. This mentality is applied also to computers as well: the mantra of a hacker is that you take a proposition and you add the phrase 'except when it's not' to the end of the proposition. For example, you could say: 'My computer lets me login when I give it the right password' . . . 'except when it doesn't', because maybe I cannot find a password that matches the hash internally (that it stores) that is equivalent as far as the computer is concerned, but in fact wildly different as far as I type it out.

*Yes, that makes sense.*

That whole mindset is not that common to be honest, though it can be taught (but not to everybody).

*I think you're right, it is a mentality. Let me share with you a somewhat embarrassing moment. I opened my first book,* Unknown Future, *with a conversation between me and a cyber security consultant named Federico, who the morning of the interview had stolen*

*€1.5 million from a bank because he was paid to do so; it was his penetration test, which succeeded. He said he had to give the money back the next day, and he was showing me his account online through his smartphone and how he got that money. The question I asked him was the embarrassing part because I'm a priest. I said, 'Federico, you're obviously very good at what you do. Why don't you just take the money and run? Why don't you just do this for a living?' And he said: 'Because I want to live in a better society.' It is the ethical consideration which is the most important. You could probably be a good cyber thief from what you know, but you choose not to do that, you're actually helping to increase security.*

I think the other side of your ethics question which is possibly even more complex is this: there is a large contingent of hackers that work for their governments.

*Right.*

They enable governments to do some very powerful things. Sometimes they agree with those things, sometimes they don't. This represents a constant conundrum for a lot of people: why would you choose to work for the Italian government or why would you choose to work for the United States government? Why would you choose to work for any government at all? At some level the hacker communities and the military industrial complex which absorbs all these hackers take very different ethical views of the world.

*Can you give me an example of that?*

I can give you an example which is possibly very close to your heart based on my understanding of some of your classes and your writings: most hackers are atheists on a level that would make Richard Dawkins seem like a moderate. This is a broad analysis. You can look at even Iran, a very religious place, so you would expect to see some religious hackers, but they just don't exist. There are a few but it's like looking for a model who eats pizza every day; they're just not there.

*That is very interesting; you're right that it's close to my heart and you have a point there.*

And this is true in Italy as well, which comes from a culture of religion. It is interesting that the ethical guidelines these hackers are creating for themselves really come from a very different place than for a lot of normal people.

*Okay. What is your own personal experience? Did you make a choice; for example, working for the NSA? That must have presented some dilemmas on occasion. If what I read in the papers is true, they are certainly getting a lot of bad press in the last year or two in terms of privacy and spying.*

Ever since Edward Snowden, of course, they have been getting very bad press, and to be honest they are not very good at public relations. It's just not something they're used to, and they hope to get better at that. I identified early on as a young adult (in my early twenties) with the hacking community. It's kind of like the cyberpunk community: there is a group of people who believe that the ongoing trend in technology is to make mankind more free and not less free, and there was a natural gravity to that equation. Therefore, we should do as much as possible to increase the access and reach of technology in all areas of our life, which is a pretty major statement of faith, in the sense that there's no guarantee this will work out. If you look at how computers were used originally, you find that they were used to count Jews to make sure that all the people went to the Holocaust that they, the Nazis, wanted to. This was an early IBM contract. That previous statement is one of faith: that this is the thrust of mankind and things are going to get better if we apply enough technology to them; not just better in an overall sense, but more full of liberty.

*I think that was the general impression in the 1990s: that computer technology was going to open up society, make us more free, more transparent and more fair. I think there was that hope.*

At some level, I think the National Security Agency aligns very strongly with that thesis, which is that they believe there are many primitive injustices in the world, which can be solved by the use of computers and other technology. At the time when I was working at the NSA, I felt strongly that that was correct. When you are part of the NSA, you see all of the barriers that have been put in place legally and operationally to make it so that they are not impinging upon people's freedoms, as much as it might seem the contrary from the papers. I read an article from the *Intercept* today on the NSA: even I can tell that it was completely out of bounds in terms of what is actually happening, but that sells papers.

> *Massimiliano Oldani [a consultant based in Rome] has always been very, very reticent to speak about things in the media, as if it is a whole other world. He says they misrepresent hackers and they misrepresent cyber security. There are fashionable things that come and go. We were speaking in very general terms but he was basically saying most of it is irrelevant: don't think that because it's on television or is in the paper that it's true.*

He is completely right! None of the open source information on news in general is ever true, once you've looked behind the scenes at the actual info. Everyone got a taste of that with WikiLeaks, right? Because you could see that when the State Department cables came out, what the United States was saying publicly and privately were very, very different.

> *Yes, and it still is. This has not passed yet. Do you want to pursue that concept referring to the NSA?*

Pursue the concept of whether or not the NSA reflects an original cyber-punkian ideal? I wouldn't really know any more because I have been out of there for so long, but I think in general there was truth to that idealistic statement that technology would make things better. I take a little bit of hope in seeing the FBI come out so strongly against encryption,[1] partially because the FBI is an example of a traditional law enforcement organization that has been accustomed to being able to read whatever it wants to read and I think the idea that individuals may someday have privacy again is a pretty hopeful concept.

*I'm trying to follow you on this. Do you really think that we will have privacy again? I don't know. Maybe you can give me your thoughts on that.*

I think our concept of privacy will change. It may be true that the government knows everyone I've ever talked to, but doesn't know what I've said to them. That may be enough.

*Okay. Do you think that we are giving up our privacy freely in order to have more convenience and more services, or is it being taken from us?*

I think it's both. The question of 'Is it being taken from you?' has to be answered by: 'Why would they want it?' And the answer is to sell you more advertising and more services from that advertising. It is interesting that a lot of the boundaries of the technology are not going to be legal or moral or even technological; they are going to be financial, based on why they would bother delving more into your information, providing you more services, if they can't get anything else from you in terms of selling you more stuff, right?

*I think Google and Amazon are pretty clear on this: that is exactly what they're saying.*

That may be the boundary. It may not have anything to do with what they are allowed to do. Maybe the way that kind of beast works is that it eats your privacy until it has run out of the ability to sell you things that come in a box. Once your money is all gone, or used as much as possible, they have no reason to sell you more shit.

*That sounds pretty clear, although it sounds depressing at the same time.*

When the money is gone, they leave you alone because they don't need to invade your privacy any more.

*A very unsettling question I asked Federico is this: seeing what he can do and what others can do, I asked him why don't people hack me? He said there is only one reason and that is because they don't want to; not because I am secure, but because they don't want to, and that is creepy.*

It is creepy especially when you understand what it means to be hacked in this day and age, where they are watching everything you do from your laptop camera, they are recording everything you say around your laptop, they are looking at all your messages and all your phone contacts, listening to this conversation. What a hacker can do is very creepy. We have a product that we have created called 'Stalker'.

*Oh no!*

This just organizes stuff you give away for free, but presents it in a really nice way to make it easy for someone to analyse your life. But the creepiest thing about it is that this product is not doing anything that your government is not already doing to you, no matter where you live. You have to assume your government knows, but has not told you they know. For example, your government knows everyone in your country who has done drugs: they just know it because they know who calls whom. And that is how you do it. Ask yourself, how do you arrange to get pot? You call a guy, he comes and finds you and he delivers pot. So they know everyone who's called him. We have not really adjusted to the government knowing that about everyone, being literally able to tell you not just who did it, but when they did it, how much they did it, where they are at any given moment at all. It is very interesting. The larger concept of what you can put together with these big data models is more and more unsettling as time goes on, and we just have to be comfortable with the fact that we are going to have fewer secrets: that part of your life can no longer be secret.

*I'm not sure I want to keep asking you questions because this is getting scary. In my original question was an ethical one: have you taken an ethical decision regarding your digital products? I assume you are using your company for good. Or are you helping tyrannical regimes to oppress their people?*

Here's the thing: it's like people asking me, 'Are you selling to a tyrannical government?' And I would say, 'I find this idea that there are good guys and bad guys to be something suitable for a six-year-old, and not suitable for an adult.' Some people act benevolently but in fact cause great harm. Other people who are malicious do great good by mistake. I also find it very judgemental. We have a long list of customers: each customer gets special attention, for we follow United States export control laws. There are things we won't do, but it's decided on a case-by-case basis at some level. We are not where some of our competitors are, but we understand our competitors and we understand the desire to have money: money is good. I think it is human nature to want to demonize people, and say these are bad people. Let's just bring up Iran again. Iran and the United States have been at each other's throat for more than twenty years. A long time. We are signing an agreement with Iran hopefully in the next couple of weeks, at which point we are going to start normalizing relations and before you know it they are going to be an ally. So, are they bad people that have turned good or is it simply that the United States is following its own best geopolitical interests and part of that is convincing your population that it is okay to hate certain countries?

*I follow you up until the last sentence here. I think you're right in that we now consider that having a relationship with Iran is in our favour, for it responds to our own interests, even though the people haven't changed radically. Before we didn't, now we do. But I didn't quite understand your last phrase.*

It's in a country's best interest to have this ability to demonize other countries and so, for example, Iranians have clerics who say 'Death to America' at the local mosque; likewise, we have our Republicans or Democrats who say 'Death to Iran' in the Congress, but I think you have to be stupid to believe either one.

*I see what you mean. You obviously are sensitive to the ethical issues and you yourself have said decisions are made on a case-by-case basis, so that means you operate with some ethical principles.*

I think we do, and I think you're forced to face it as a hacker probably more than in some careers. There is a girl I'm dating who deals with immigrants. She has spent her whole life doing good for others, she is helping people who are downtrodden in the most official way. She is helping them one by one, but she is still helping them. For me, hackers are making big changes: you're manipulating a giant geopolitical machine and what you do may change things very drastically or not at all, or may change things in one direction when you hope for a different direction, and you have to confront these ramifications on a daily basis. One does live by a set of ethics and honestly I think for a lot of people it's just an emotional thing they never rationalize.

*Yes, I think you're right.*

So Darfur is emotional for a lot of people. They think that Sudan must be full of evil people, but the reality is that you might be selling to the part of the Sudanese government that is tracking down Al Qaeda, who are also pretty evil. So, maybe that was a good win. Maybe they got three Al Qaeda guys for every one innocent guy. I don't know what your best equation here would be.

*You do business with these people, don't you? You are a global company?*

We do, but not with everybody. Actually, our business is only 30 per cent international, a lot of it is steel manufacturers or people you would not consider a part of any kind of controversy.

*When you need to reflect on the ethical parts, is there a committee or is it just yourself making decisions?*

We have vice-presidents in the company and so the company makes the decisions internally. Depending on the decision, different people will be involved but there is no formalized ethics committee within the company. We are a little too small for that right now.

*Okay, but it seems you take it very seriously.*

I think we do take it seriously, because we want to be smart about it.

*As I mentioned in one of my emails, almost everyone in the first book told me that cyber security is the number-one priority for the next five to ten years of technological development. Can you speak to that?*

Everyone says that, but then everyone would like to make it everyone else's problem. Let me give you an example. Right now, there is the regulatory framework, which although it comes from the EU is signed by the United States. It is called 'Wassenaar', which is traditionally an arms control framework, and so it controls nuclear weapons and other devices that can be used negatively. They recently added to that the ability to control software – specifically, intrusion software – and the reason they did that ostensibly was because activists and dissidents around the world were getting hacked into and found and then killed.

*Wow.*

Yep, it happened in Ethiopia and a few other places as well that people find unsavoury. They would love to regulate all software produced in the whole world so that nobody could build hacking software and then sell it to the Ethiopian government. This is their stated aim. The downside of that is that regulating software (which is essentially an idea encoded on the page) is an extremely painful, difficult and onerous thing to do for all parties involved. In order to prevent people like activists from getting hacked, they then have to regulate the behaviour of everyone else on the planet, because they are hoping to make it someone else's problem. One simply uses some decent operational security and has activists running the latest iPhone and then they won't get hacked; but they will probably get killed some other way, via some other mechanism, because the government is still going to find them and kill them. But this is why they would much rather make it everyone else's problem. So, when people say, 'Oh, cyber security is the biggest issue in the world', everyone means that, but what they really mean is that they hope someone else deals with it and not me, because I don't want to pay for it. It is similar to global warming, pollution, the typical tragedy which is common but now written on the electronic page.

*But you are obviously doing something about it. Is this not why you started your company?*

In a sense we perform a service like a dentist. We provide you the necessary hygiene that will allow you to keep your teeth longer. We're not going to guarantee that if you go out and eat a bunch of sugary foods your teeth aren't all going to fall out anyway.

*Okay.*

It is a valuable enterprise. We take money for our efforts, but we're certainly not claiming to save the world here. That's not what cyber security does.

*Many corporations I think see cyber security as a priority. For example, I interviewed the chief information officer for PricewaterhouseCoopers, which is a large auditing firm, and he said absolutely priority number one through number ten is data confidentiality. PwC cannot allow any breach in confidentiality of their data. They cannot allow themselves to get hacked, they cannot allow their data to get compromised.*

When they say priority, what they mean is fear. It is not priority. Priority is different. We all have two things in our heads: priority and fear. You can tell what your priorities are by looking at where you place your money, and your time and your energy. In terms of your fears, they don't necessarily drive your behaviour, but you still have them; fears are emotions that we cannot deny, but that does not mean you act on them. So, I would say yes, PwC greatly fears that it may suffer a major leak event, where all their data gets leaked, and that would hurt the company. It would be devastating. In reality, though, I don't know how bad it would be, there is no way of knowing. However, just because they say that does not mean that they are not going to spend more money every month on advertising than they spend all year on cyber security.

*I just don't know, I would have to check.*

Not just money, but also thought, because a lot of people put money on stuff just to cover their ass. They don't really care if it works or doesn't work, but the fact that they spent money on it makes them feel better. That is a really big deal in security because we have all collectively, as a society, spent a lot of money on technology for cyber security that we realize does not work.

*No, you're kidding me.*

Look at your own email! At the bottom of your messages it says, 'Secured by Avast', which is a company from the Czech Republic working with antiviruses. They are probably reading your emails, because you've already installed it for them.

*It's free!*

So, how do you think they're paying for it? Furthermore, the reality is that, even if it were working perfectly, it would not protect you at all, and those are the kinds of things we spend a lot of money on.

*That is really unsettling.*

We always knew it was true and we really didn't care until things started happening to us. It's just like global warming: we all knew it was true, we just didn't care. I live in South Beach in Florida just three feet above flood level. If we have a real bad storm I'm twelve feet underwater and then I'm going to start caring.

*It is kind of depressing to hear you say that because it means that as a race we are kind of stupid.*

We are totally stupid as a species.

*So, is there any hope?*

Get off the planet, I guess, would be my hope. Seems the only long-term solution.

*What do you see within the next five to ten years in your line of work? No one has a crystal ball, but we can analyse trends.*

I would say there is currently an extremely interesting war going on between governments and industries, all over the planet. Governments are very much used to having a monopoly on power, but they are unable to protect corporations from cyber security losses, especially cyber security losses that are originated from other governments. Most corporations are now very international, and the more the internet comes into play, the more international they become and the less they want to depend on or allow any one government to control them. I think we're going to see a lot more government alliances that collectively provide what a single government used to provide, which I guess in Europe would be the EU; but of course the EU and the United States are bound together in more ways than people realize: financially, with security agreements, conjoined laws and societies in arrangements and just simple relations. They are getting more and more joined into one thing, even though ostensibly they are two separate areas. If you look forward I think you're going to start seeing the flux of power change: governments used to have all the power, but I think they are getting less and less powerful overall, which may or may not be a good thing (because I don't think corporations are beacons of hope in any way).

*Corporations don't seem to be the most altruistic organizations.*

They are explicitly not so. But they may bring other benefits. What I've learnt is not to judge things, but simply to observe them. We can see this in the schisms between very basic primitive issues: are corporations required to give up information to any government that asks for it?

*It seems to me that is the tendency.*

I don't know. People are fighting it. They are saying: 'No, not only are we not going to give you such information, we are not even going to hold it for you. That information does not exist wherever you want it.' Right now, the United States and Microsoft are coming to blows over whether or not the United States can ask for information that Microsoft was storing in Ireland. Microsoft is saying, 'You can't ask for it; you have to ask the Irish for it and they might be able to get it for you but we don't know.' The US is saying, 'No – you are an American company, you need to get whatever we want you to get, from wherever your servers are in the world.'

These kinds of battles are planetary, they go up and down the ideological scheme of things. It starts there, and goes to computer security with the whole Wassenaar stuff. You could go to encryption, which we saw yesterday in the US Senate where they discussed whether or not companies should be forced to back-door their encryption for the US government. The answer was, 'No, we are not doing that, you guys can leave us alone.'

*But do you think those companies do it for their own purposes?*

I think they have different value systems than the governments do. They look at this as a trade: if governments are not going to protect us, why should we give them the information?

*Why would they say the government is not protecting them?*

For example, Google cannot fend off the Chinese hackers, and the government won't do anything about it. The government is not going to go to war with China over Google getting hacked, and that's pretty much what Google needs them to do.

*That is certainly scary.*

Look at Sony Pictures. They almost went away because they were hacked, and they were about to go down, that could have been the end of Sony Pictures and it probably will result in significant job losses. It has fundamentally changed what media companies in the United States are willing to publish.

*Why didn't they go down?*

At the last minute the United States government stepped in, and told the North Koreans to knock it off. But they weren't going to, because it's not their problem. Protecting Sony Entertainment shareholders is not the US government's problem, but it has become their problem to protect Sony.

*I have the feeling that the United States would be willing to do many things to protect Google because it's a source of information, a unique source of information.*

Yes, many things, but what they also want to do is to require Google to do things for them that Google does not want to do, like, for example, back-dooring Google's phones, based on the Android software. Google wants to retain their 80 per cent dominance of the cell phone market, not allowing a Chinese operating system to come in and displace them. And the minute they cooperate too much with the United States government, that is what's going to happen.

*Why?*

Because the Chinese will market a system which is not back-doored and the Europeans will buy it.

*That does make sense. It is a very delicate balance.*

It used to be a delicate balance. Now I feel like it's a free-for-all, like one of those battles between cats and dogs pictured in the comic books. On every front there is conflict, everywhere you look there is conflict.

*Is there such a thing as foolproof software or could there be?*

No. There is no need to equivocate about that: it's plain and simple.

*Do you think that people are trying to work to create 100 per cent foolproof software?*

No, not really.

*So what is our option – to just live with vulnerabilities?*

I think a little insecurity is good for you, it provides you some necessary freedom. Perfect security is its own kind of tyranny. If you were to come up with this software, it might restrict you as well. I don't think it will ever happen, so it's like asking if a giant unicorn is going to come down from the sky and fart rainbows at me.

*In terms of your own company, when I asked you what you see as a trend for the next five to ten years, would you offer another example? Would there be something else you could identify as a trend?*

Everyone on earth is saying that the trend is mobile, but realistically the trend is towards artificial intelligence: this year, weirdly enough [2015], we have seen that artificial intelligence has really come into its own and it has happened without people realizing it. One of the key examples is this: I give you a picture and you as an adult human being can tell me what's in that picture; but until this year, we were not able to achieve that with computers. Now all of a sudden it's easy. You can go to Google Photos and you can type in, 'Show me all the dogs' and it will show you every picture you ever took of a dog. That level of ability to solve problems is a big inflection point. I can have my kids talk to Amazon's 'Echo' speaker, based on Amazon's artificial intelligence: they can say anything they want to and it will respond to them, it will kind of 'understand' them. The idea of what does it mean to have understanding and consciousness is changing, and has changed. We just don't realize it yet.

*That is very philosophical.*

Have you read the book *Blind Sight*?

*No, I have not.*

*Blind Sight* [by Peter Watts] is one of the best science-fiction books written recently, although it's very depressing, as some of the best science fiction tends to be. It deals with the differentiation between consciousness and intelligence. The concept is rather simple: there are many people who are blind, and yet if you throw an object at them, they will still catch it, and that is because they are operating without using their consciousness. They cannot actually see the object, but their body can still catch it.

*Does that actually work?*

Yes, it is real. That is where the name of the book comes from. The premise of the book is that if you examine the fact that consciousness does not have to exist for intelligence to exist, at that point things quickly get very scary. This is what you are about to see with artificial intelligence.

*What would you identify as some of the more important projects for AI?*

There are not that many AI projects in the world, nor will there be for a while. I think Amazon has one. Google has one. Apple will end up having one.

*Are you familiar with IBM's Watson?*

Yes, but that is really a small piece of AI. If it can't drive your car, it's not an AI.

*I saw Watson defeat the reigning champions of* Jeopardy!

Yes, it can, but it's almost a parlour trick when compared to recognizing a dog in a picture.

*Is it really that difficult to recognize a dog in a picture?*

Yep, it is that hard.

*Once AI works, it disappears and becomes seamless.*

Maybe we're going to end up calling it computation without realizing what we've built.

*Do you think there are inherent dangers with something like that?*

I think there are opportunities. I think every generation has to scale the ideas that come across every new generation. Perhaps my grandfather's generation had to come up with an answer to mutually assured destruction, because we had the capability of destroying everyone else on the whole planet, and we're going to have to live like that. And here we are, everyone has nukes buried all over the planet and we've all grown accustomed to it. AI is a similar idea. It is hard to get to grips with it at first, it's scary and then eventually it's everywhere.

*Do you see your company developing along AI lines?*

We are a bit small, we do advanced things in some places. We are not in the AI market.

*I think we are already seeing intelligence without consciousness. I often ask people if they think their jobs will be taken over by machines. In your case, what do you think?*

I think many people's jobs are going to be taken over by machines. What's more, most of those people are probably doing their jobs poorly, and so I'm glad that their jobs will be taken over by machines.

*Do you think it could happen to the founder of Immunity?*

My job is pretty weird right now, so probably. People could put a Twitter box up and pretend it's me.

*Is there a thought or a final consideration you would like to make?*

I don't know that I can give people advice, but my thought is that you can have major reactions to everything, but I feel like the more you just embrace the new subtlety of the world, the better off you are going to be. I like to think that the original model of technology, as making us free, is still true.

*I'm optimistic also. I think in the long run, we will learn how to use the technology in a more human way. Right now we're on the learning curve, we are still infatuated with technology.*

Thank you for conversing with me.

9.

# *Journalism*

## (*Christopher Altieri*)

Vatican Radio transmits the Pope's voice all over the world. It represents a valuable service to Catholics and non-Catholics alike. Having known Christopher Altieri from a mutual experience at IES (a study abroad programme in Rome), I asked him if he would be willing to discuss with me the impact that digital technology is having on journalism. He eagerly accepted. We met at his desk at Vatican Radio.

Perhaps it is overly simplistic to suggest that digital technology has replaced the need for journalists and reporters. I think quite the opposite is true: now, more than ever, we are experiencing an information overload, saturated with data. The journalist is the one who makes sense out of it all. Christopher is also persuaded of the timeliness of his profession, as is made clear by the following conversation.

* * *

*Who is Christopher Altieri?*

Who is Christopher Altieri? He's a guy who was born in lower Manhattan in a Catholic hospital that no longer exists, St Vincent's. Grew up in southwestern Connecticut in a small city called Stamford, not as small now as it was when I lived there. Went to Fordham University for two years after Fairfield Prep.

*So, you're a Jesuit product?*

Yes. As our Italian hosts would say, *Gesuita mancato.*

*Did you think about a vocation?*

Yes, I did. I came here to Rome in 1997, after my sophomore year at Fordham University. There I'd fallen under the influence of a professor who'd done his doctorate at the Gregorian University.

*[chuckles] Under the influence.*

Yes. In perhaps more ways than one, you know. Quite seriously though, I came to study philosophy at the Greg and to live the intellectual life in the service of the Church. At the same time I was seeking something of a geographic cure to a broken heart and discerning a vocation to the religious life and specifically to the Jesuits because I had, for as long as I could remember, been under their care for my education. I grew up in a household that had great admiration for and acknowledged a great debt of gratitude to the Jesuits who had been my father's teachers in secondary school, in college and at law school. He's a Fairfield Prep grad himself, Georgetown Undergraduate and Law School.

*That's vintage Jesuit material.*

College in 1968, law school in 1971. For folks who keep track of these things, it was before, or right at the beginning of, the silly season, as it were. His was still a sort of old-school formation, and I felt the same debt of gratitude for the formation I had received. I thought it would be ungrateful of me not to consider it – I mean a vocation – and I have to say I found the idea – and I still find to this day – the idea of a life dedicated to Ignation spirituality very attractive. So what happened? What happened was I skipped class one day and was at the tomb of Saint Ignatius praying and asked him . . .

*At the Gesù Church?*

At the Gesù, yes. I was there asking Ignatius, 'Do you want me?' The thing was, I had an appointment with a friend. As it turns out, it was the same professor who had encouraged me to come to Rome and to study at the Greg. I walked back to the university and met him at the front door, the three arches that are in the main building at Piazza della Pilota. We went in, walked up the stairs, went up to the first floor where the library is. As we were going up, there was a girl coming down the stairs and I had to say, she was just the most beautiful thing I'd ever seen [laughs]. You know, in those days, anyway, that sort of creature was not exactly 'run-of-the-mill' in the halls of the Gregorian.[1] So my friend said to me, 'Good-looking girl,' or something to that effect, and without missing a beat I said, 'Doc, I'm gonna marry that girl,' and I did.

*How about that.*

I wrote about this recently, in connection with the feast of Saint Ignatius. Roughly, I said that one of the characteristics, perhaps *the* characteristic – the defining characteristic of Ignatian spirituality – is freedom. People have said, and I think it's right to say, 'Oh, look. Saint Ignatius answered your prayer.' No self-respecting author could pen this story and sell it as a work of fiction. It's just not plausible. The way that he answered was typically Ignatian. By his intercession, I'm quite convinced, Ignatius obtained for me the grace of seeing the choice that was before me, and he turned the question around on me. He said, 'Sure, I'll have you if you are interested. I'm here, my men are here, but what do you want? Here is the good, right before you, pursue that, and love her well if she will have you, we are still here but ask yourself, what do you want? And if you ask yourself honestly and answer truly, then Christ our saviour will be your good Lord in all things.' I was not thinking this at the time, I was thinking what I needed to do to get her number.

Let's be very clear about this, this is all reflection and hindsight and successive understanding, but it also can give you an idea of the influence that Ignatius and his Company have had on me and continue to have.

*What that exemplifies is typically Ignatian and it's called discernment, the discernment of the spirits. Ignatius' Spiritual Exercises are really for that and, on the other hand, you were lucky because she also found you attractive [laughs].*

Well, I mean it took some work. I have to say there are probably some jurisdictions in which my courtship would have technically qualified as stalking, but all is well that ends well [laughs].

*You didn't give up, in other words.*

No, I wore her down, as they say.

*Okay, now when you were studying here, were you in a Collegio?*

No, I was a layman living on my own and enjoying that very, very much.

*So did you end up getting a degree from the Gregorian?*

Yes. I defended my doctorate in 2010, in philosophy. It took a while to get the project approved and then done for reasons that we don't need to go into – the typical vicissitudes of post-graduate work. I have prepared it for publication, and it is now out.[2]

*You were working then at the Radio Vaticana?*

Yes, I've been here since 2005.

*I want to talk about how the digital revolution is influencing your work, and it's interesting because you've lived in the transitional period. Everything is digital now.*

Yes, and what little journalism I did before the gig here I'd done as a print journalist, doing special reporting in columns and things like that for a local paper back home. We're now almost at the point where people of adolescent age (if not young adults) are surprised to learn that printed newspapers were actually at one time delivered daily to people at their homes and work places. That's an outlandish tale. That has been one change that I've observed: it hasn't affected us directly in any hugely important way here at the radio. But the change in print media has affected all of us. There are organic changes happening across media, old and new; and radio's interesting because it straddles the old and new media.

*Yes, many people thought radio would disappear with the internet, and I think just the opposite is true, isn't it?*

Absolutely, traditional radiophonic broadcast technology is perhaps not as prevalent as it used to be, yet there are still plenty of places in the world where it is used. One of the things the Vatican Radio does is try to reach those places via traditional radio broadcast. Remember, you can drop a small FM transmitter somewhere and you can start to piggy-back and send signals in the middle of a refugee camp in Africa.

*You guys do that?*

We do that, yes (we broadcast with a view to reaching places like refugee camps and other places that are otherwise not 'on the radar' of traditional media outlets). That's the mission statement. It says that right on the front door: to be the voice of the Holy Father and of the Church in the world. We are going out to the peripheries, as Pope Francis says. To borrow a turn of phrase from the hipsters, we could say, 'We were doing it before it was cool.'

*Yes.*

And we continue to do it. But how has technology affected our work? We go out via the internet, we internet stream, we have a Vatican radio app. When we do special programming for Papal events, we go out on literally every multimedia platform available. As an institution, we are still wrestling with how best to use the new technologies in service of the mission. We are also trying to figure out how to tailor our 'product' to the new technologies to take best advantage of them. That's not something that you do overnight, and it's not something that you can do only piecemeal. And it's also not something you can do effectively simply around a committee table with people making decisions and then implementing them without having a good deal of trial and error. *Dal tavolino*, as our Italian friends say.

*Let me zap you with something you may not have prepared. What you're talking about is a two-edged sword in terms of using new media, but for a very old purpose, and that is to have the voice of the Holy Father heard throughout the earth. This Holy Father is unique in many ways. One of the things that I think has probably given you headaches is the homilies at Santa Marta. Now everybody wants to hear those, but as Father Lombardi says, 'They weren't intended to be heard by everyone.' When he gives those homilies he's talking to forty people at his mass. Yet everybody wants to hear what he's saying, and you have the technology to do that. So, has there been discussion about it?*

Sure, there has been discussion about it, and the Holy Father has said that he does not want a full transcription or a full audio recording released to the public.

*I bet there's an audience for it.*

Oh, I'm sure there is an audience for it. But, you know, he's the one over there calling the shots, not me. It has been something that we've had to wrestle with. The Holy Father knows that and he's okay with it, this is the way he wants to do it and that's his decision and there we have it. The Holy Father's remarks following the readings of the day are recorded.

*Because he spends a lot of time thinking about those.*

Yes, you can tell. They're recorded and a transcript is made. Then some of the best bits from the transcript are chosen and a summary is written up (usually in Italian), and is transmitted on the semi-official 2.00 p.m. news broadcast: it's either twenty minutes or twenty-five minutes, and it's Vatican Radio's take on the news and events of the day and the week. The Holy Father's remarks at morning mass are a part of that. We in the English section get the finished product, the summary with the poll quotes mid-morning. Usually not before 10.30 or sometimes even 11.00 because by the time we get it, it's not only that they've produced it but it's been sent up to the Secretary of State, they've vetted it and sent it back. And we go with it.

Then, we translate it into English and present it. Obviously the parts of it that aren't in quotes (the summary part) don't always get reproduced verbatim, simply due to journalistic expediency and cultural differences in the way that native English speakers and Anglophones in general from whatever part of the world process and consume information. The quotes remain the quotes and we do those and present them fully, and that's what people get. Sometimes it's a challenge to translate. There's some preparation, they're based on the readings and they're short, pithy exhortations. He's not doing theology in these speeches, whereas with Pope Benedict you could very often get that – like when he would speak to the International Theological Commission, you could get an impromptu theological treatise from Benedict. I remember one time, early on in his pontificate, he put aside his prepared remarks and he spoke in printable paragraphs [laughs]. You could just cut them and take them downstairs and run them.

*Pope Benedict was a professor, after all.*

He had his citations in there, I'm talking chapter and verse, but not just chapter and verse. 'As Irenaeus says in letter 126,' or something like that. I remember (just for kicks, one time) going to see whether his citations actually checked out. I don't remember one of them not being pretty much where he said it was. Francis doesn't do that. No, I don't think that there are a dozen people on the planet who could do that. He's not interested in it. He's interested in giving pithy reflections on the readings and a moral exhortation, which by the way is what every priest who says a 7.00 a.m. mass and gives a short homily is interested in doing. That tells us where he lives. He's a pastor, and he's very often talking to the people who are there, and he's tailoring his remarks to them.

Sometimes he's got his mind on something, and the intellectual work that he's doing as he's making his remarks is that of seeing how the readings of the day inform his thinking about something in particular that he's got on his mind. When that's the case it's not always instantly clear what the connections are. I've never once (having sat there and thought about it) not come to see it, but it doesn't happen right away.

*The connection you mean?*

Yes, and sometimes it's quite profound and, you know, it's jaw-dropping and it can keep you up at night thinking about it. How did he see that, how did he get there? He had gone to the bottom of the ocean obviously.

*Oftentimes, his comments will be on the nightly news. I mean national and international news.*

Sure.

*And you're right, they're not 'off the cuff' but they are spontaneous, he's not reading a text. That must be a little disconcerting.*

I'll tell you how. We're under enormous deadline pressure. There is a twenty-four-hour news cycle. That beast must be fed. The beast is hungry.

*It must be terrible. Because you guys go to bed, but you wake up the next day and start it all again.*

The radio under normal operating parameters does shut down and open up again. We get a break, but the news doesn't take a break. That is one of the things I think that we explore and how we can be constantly present in a way that is a real service without burning ourselves out on the hamster wheel. There's no perfect solution and it's not an easy thing, there's no quick fix, but we are under enormous deadline pressure. Nowhere do you feel the pressure or the tension between getting the story out and getting the story right than you do when you're translating a guy who is wrestling with complex, vexing moral questions or gigantic questions of Church governance . . . thinking in Spanish, speaking in Italian, offering remarks that are spontaneous, and have the character and flavour of spontaneity. How do you preserve that stylistically in your translation so it doesn't look like a printed paragraph, or a book, or a prepared speech, and at the same time get it right? What if you miss something? Very early on in his pontificate he said the mass in the Sistine Chapel with the cardinal electors on the day after his election. He used this turn of phrase (I think he must have been thinking in Spanish), talking about the Church, and how in everything that the Church does, she must live and breathe charity. Otherwise, she's not who she is. Without that, he said, the Church could end up becoming '*un NGO pietoso*'.

*I remember that.*

We translated that as the 'Church could become a pitiful NGO'. I remember thinking about that and actually thinking that what he was saying was 'pitiful' in the English sense – it could be a poor copy of an NGO. First of all, there's nothing to say that that ambiguity isn't present in the language that he chose. So, which one do you go with? Turns out, he didn't mean it that way. That's just one example.

*Yes, that's a tough one.*

This happens every day, sometimes several times a day. We can't spend for ever arguing about it. We've got to make a decision and we've got to go with what we go with, and sometimes that becomes *the* translation. *The New York Times* or some other newspaper picks up on what we said and they go with it. Another example was having to translate *viscido*. We translated that as 'smarmy'. When it's something like that, it's cute and it's fun and it's a feather in our cap, because we were the ones who got it right. But it's not always that inconsequential. There are times when it can be very, very important and a matter of considerable significance to get it right, and to do that under the kind of pressure that we have to be out there, not just to be first past the post with our write-up of the Sunday Angelus message. Think of the masses out there, because if we don't do it, nobody has it.

*Exactly. Let me go back to the issue of technology. Marshall McLuhan said, 'The medium is the message,' remember?*[3]

'The medium is the message.' If you'd asked me five years ago what McLuhan meant when he said that, I could've told you that I figured it out, I know what he means. With the multiplication of media and the various and sundry ways that we communicate, I'm nowhere near as confident now as I used to be in my understanding of what McLuhan meant. But I think he was certainly after something like this. He was tapping into something that the classical rhetoricians understood. Cicero said that rhetoric is one great art composed of five lesser arts: *inventio, dispositio, elocutio, memoria* and *actio*. The means by which a message is conveyed determines the mode in which the message is transmitted and that medium determines also the content of the message. I was thinking about this just last night actually, because I was watching Ken Burns' *Civil War*, and listening to Derek Jacobi read copy from an English journalist describing Abraham Lincoln. He was giving a physical description of Abraham Lincoln. I couldn't get the famous black and white photograph of Lincoln, standing in profile, out of my head. It was actually causing me consternation. The photograph does not convey to the receiving subject the same thing that the written description does. That again is changed when you get someone reading the copy of the description. And when it's Derek Jacobi, you have to deal with that also. The question is forced upon us: they're all talking about Abraham Lincoln and they're all describing in some way, *mutatis mutandis*, Abraham Lincoln. What does that mean? If I can wax philosophical here for a second, what are the implications of that for our ability to get 'at the thing itself'?

*Good question.*

The multiplication of media presses upon us with an ever more palpable urgency that great and ancient philosophical question. I don't have an answer to it.

*Especially in your profession, there is no easy answer to that. Your profession has also been accused of lying.*

Sure.

*Because you're always representing a perspective. There's no such thing as a non-perspective perspective. You're always choosing data and this is what makes a great journalist. There are good journalists and then there are bad ones, but you're always giving a perspective, aren't you?*

Here's one of the advantages of radio though, as long as we're talking about the strengths and weaknesses of the various media that we have. Yes, sure, voice recordings can be faked. They can be altered, but it's always good to have a recording of a guy saying what you said he said. The reason we have records is that you never know when it might be important to check a source, and an audio recording enables you to do that. These are things that can be manipulated, but if all you have is a transcription or a stenographer's report of something, then that's all you have. An audio recording is different. A video tape with audio over it is the same thing. Now you've got not just a clerk's report of who was present at the meeting, but you can see who was there.

I think you'll see a lot more media outlets in general recording things and then archiving them so that multiple sources of the same recording can be made available immediately to the public. Let me put on the professor's hat for a minute. You know this because you're a teacher yourself. To say that students have a difficult time following an extended argument is perhaps to put one's self in the running for understatement of the year [laughs]. They cannot process more than 140 characters at a time.

*You're right. Yes. It is sad, though.*

So what do we do? We meet them where they are, as Francis says. In this case it's not 140 characters, although he's doing that. He's tweeting the message. He's making use of that particular medium, that particular platform. Fantastic. Then you've got our write-up of his story and it's 150 or 200 words, you can flash through it for maybe forty-five seconds, maybe a minute of audio that you can listen to. You can listen to it live, you can listen to it streamed, you can download the MP3 and listen to it on your iPad or your Android device at your leisure. You can get the extended write-up, you can have the full text of the speech right underneath the 150-word summary, and if it's an article that you've done, an interview with some Vatican official about topic x, y, z that's in the news. You've got your full quotes in the short write-up, let's say no more than 1,000 words. Who reads 1,000 words anyway? Might as well be *War and Peace*. You've got that and then you've got the full transcript, you've got the one-minute audio report that's a very synthetic presentation with maybe the one poll quote, the 'money' quote from the interview. Then you can stick the extended interview right next to that and right under it so that somebody can go to it if they want. Journalists can use that. The permutations of this are not infinite, but they're manifold. We journalists are still getting our heads around how to use these things properly, and more so the consumer of the information, the public that we're wanting to reach.

*Yes, and it's just happening so quickly. The technologies that you're describing are free and ubiquitous now. Whereas before, maybe ten years ago, they would cost a million dollars and only three or four outlets had them. Now it's just completely pervasive.*

And saturated.

*This brings me to an issue that I think is extremely rich: anybody can be a journalist today. You get a blog, you're out there and some people would say, 'Well, the internet is going to be the demise of professional journalism.' I think just the opposite has happened. I think because everybody is a journalist now, I need to go to trustworthy sources. That's why I think that Vatican Radio is not only surviving but it's thriving.*

One would hope. One would hope that that would be the case. These problems are of such tremendous complexity and they are cultural as well as technological. You can see the crossroads between humanity and technology as they play with each other in so many different ways. I'm almost embarrassed for a starting point, but let me try it this way. Very often journalists who are not on the religion beat will be assigned to cover religion stories. Rarely in my life have I met a journalist who, *qua* journalist, didn't want to get the story right.

*That's healthy.*

But here's the thing: you hear a lot of complaints, especially in the circles in which you and I run (i.e., the professional Catholic circles), such as, 'Oh, the mainstream media, they're out to get us, they hate us and they don't care.' Okay, first of all yes, there are some people out there who have abandoned even the pretence of journalistic integrity and they are engaged in ideological warfare. Those people exist today, they have always existed, they will always exist, and frankly, sometimes, when you are part of an organization like the Catholic Church, you're going to have to do business with those people. Let's put them aside for a second, acknowledge their existence and move on.

*It'd be naive not to.*

What you have here is a situation where reporters want to get the story right but they don't know how to. They're not themselves religious in any meaningful sense of the word. Even if they are Catholics, they're very often either ex-Catholics or poorly catechized Catholics, and they've been brought up on this idea of the fourth estate as a muck-raking, iconoclastic proclaimer of truth to power – that in order to do its job it must be suspicious of, if not actively distrustful of, official sources of information. So they don't come to us. Then these people complain, 'Why didn't they tell us?' I respond by saying, 'Well, it was right there. You wanted the whole speech? Here's what he said.' The Pope is misquoted and that misquote goes around the world 427,000 times.

*It's happened.*

We're sitting there going, 'Well, he didn't say what you said he said. That's what the problem is. He didn't say that thing you said he said.' Now, any other reporter (not only the reporter but also his editor who cleared it) who misquoted a public figure of the salience and importance of a Roman Pontiff on a question of such importance as Catholic teaching involving (for example) abortion and whether it's still a sin, would be seeking other employment.

*Be out of a job.*

Not just out of a job. That person would be seeking other work in an entirely different field because he could not get hired as a journalist anywhere in the world. Yet none of that happens. And there's not even a retraction printed. There is a limit to what we can do, but that doesn't mean that we don't have to do it. The Catholic Church tends to take the long view on these things. That's why having the record and preserving it is important. We also believe that the truth will prevail. So there's reason to be, if not sanguine, then cautiously optimistic and at least to have a picture of where we want to go, and we can start to develop an idea of how to get there. I've said some things thus far that could suggest that I'm not terribly sanguine about the future of communications generally and of our ability to be effective bearers of the message to live up to the ideal of our patron, St Gabriel, Archangel.

*I don't think you're pessimistic. I think you're realistic.*

Let me give you an example from where the rubber meets the road, about why. I like this example. Let's go back to Pope Francis. Very often he has spoken in unguarded moments. I think that he has been the one to deliberately let his guard down and knows what he's doing. He's been criticized by people (especially people within Church circles) for imprecision and for carelessness in choice of words. Whether and to what extent those criticisms are on target is something that I think the Holy Father, and his confessor, and the Almighty can work out amongst themselves. I will say this: I've never, ever in my entire life met a Jesuit who does not say exactly what he means, and mean exactly what he says. Watch what happens when the Pope says something like, 'Who am I to judge?'

*For example.*

For example. The mainstream secular media pick up on it. They run it around the world. And all of a sudden, every single Catholic apologist, theologian, Catholic philosopher, Catholic journalist is out there explaining it. You know what? Very often they are getting phone calls from the news desk at the local news station. They are getting a chance to explain it and they are getting a chance to comment on it. All of a sudden people are talking about what the Catholic Church really teaches and believes and understands, in public, and they've got an audience for it. That sounds pretty great to me. It's messy, but it also tells us that when he says, 'Go out and make a mess' . . .

*He did say that.*

. . . that's what he means. Mission accomplished. Perhaps, when he goes and decides to make a mess himself, he could think a little bit more about the people who have to clean it up (though I'd add that their reward will doubtless be great in heaven, and as for us, well, we do get a cheque at the end of the month, so . . .).

*Let's not stray too far from this, look at what happened yesterday: Pope Francis actually went to an eye-glass store at Piazza del Popolo to get his glasses fixed. The event caught everyone by surprise and was completely unexpected. Even the owner had no idea this was going to happen. Almost immediately a crowd formed around the shop.*[4]

He went to get his glasses fixed. Why is that such a big deal? It's a big deal because the Pope went out. He doesn't usually do that. It's all over the media. I can't afford glasses from that shop [laughs]. Not on my salary. That's the guy that he has been using when he's in Rome, fine. I like the idea that he went out there and I would actually like to see more of that. It is impossible for the Pope to get out from under the bubble. All that happens is that the bubble moves with him.

*Here's my interest with you and the future of journalism and communication. Someone with a smartphone was more powerful than the entire Vatican Radio at that moment because you guys weren't there.*

No, we weren't there. No, no, we didn't know where he was going.

*Right. But some jerk with an iPhone is capturing this and that is now going to be on ANSA, AP, CNN . . . all the major media outlets.*

Sure and great. He was there. My collegue Charles Collins once happened to be walking by the Synod Hall during the last Synod session. The Pope was on his way in, he pulled out his phone and he filmed it and it was this great close-up that we had of him, this raw video, as we say, raw 'footage'. Is the guy who films the Pope buying his new lenses a journalist?

*No. I don't think so.*

Is the blogger – someone who makes his living blogging and sometimes breaks stories – is he a journalist? Where do you draw the line?

*I'm less sure about that. I see your point. But for example, the* Huffington Post *is a blog but it's probably more important than most newspapers.*

In terms of how many people they reach and who gets their news there, they very often have some well above average work, that's true. They have an editorial line at the *Huffington Post*. It's not one that I always agree with. But you know what? When you're an outfit like that, you get to have an editorial line. I bring this up not just as an aside, I think it's actually one of the central points and one of the things you can use to determine whether what you are dealing with is journalistic and whether what you're looking at is journalism properly understood; whether the outlet or the outfit is, properly speaking, journalistic. Do they have an editorial line and do they have reporting standards? If they're telling you you're getting straight reportage, is that what you're getting? Do they have an editorial line that is, if not explicit, then transparent? Those would be *indicia* that I would look for across media and across platforms to figure out what journalism is in the second decade of the twenty-first century. You're the first person who's asked me this.

*Yes. That's not easy. Take, for example, footage that we see of ISIS blowing things up: often it is not true. I find this fascinating. I've spoken with people in the Rai [Italian television station] who have presented video clips as authentic, for example, in Palmyra, 'they're blowing this up'. They're not.*[5] *They're faking the video and then they're selling artefacts on the black market and making a lot of money. The Italian police are in the first line of defence on this. That's how we know.*

Yes, and one of the things that happens with social media especially is that someone produces a video that is either completely cut from whole cloth and fake, or extremely doctored, or shot in a way to make it look like something that hasn't actually happened. Goes up on social media, makes the rounds, there's outrage because that's sort of the default, the go-to reaction on these things. But did it happen? In all justice, as journalistic outlets we have to be concerned about whether what we say happened, actually happened.

*What do you mean when you say 'we'?*

I mean people in general.

*Yes, humanity.*

How does this affect us as journalists? It goes back to that tension between getting a story right and getting a story out there. Okay, we've got the footage that's making the rounds and let's say that you're a television news station and you ask: 'Are we going to run this on the news?'

*For example, did you hear that reporter Alison Parker and cameraman Adam Ward of WDBJ-TV were fatally shot during a live interview on 26 August 2015, in Moneta, Virginia? I think the event shocked the country.*

That episode has the added complication of being basically a 'snuff film'.

*Absolutely, that's the correct term.*

Why are they putting this out there? If I had been in charge of the newsroom, I would not have run it.

*I bet you that temptation was just too great.*

I bet you that I would have made that decision and the next day I would have been looking for other work. Because the people who are in charge of the people who are making the journalistic decisions and editorial decisions on the news are in it to make a buck, and they're losing ad revenue by the second, every second that you're not running that thing on a loop.

*I'm just glad I've found somebody who still grapples with the issue, because I think in the States they saw this as a gold mine.*

Yes, but you see this in reportage of all sorts of stories. You end up with twenty-four-hour coverage of, for example, that horrible plane accident.

*Which one?*

The Malaysia Airlines plane that went missing in the Pacific (or so we think).

*And we still don't know where it is. It's a mystery.*

And we still don't know. But do you remember? It was in the news for twenty-four hours a day, seven days a week, constantly. Someone's grandmother's cousin's sister's friend of a former roommate maybe one time had lunch with the guy who was going to be on the plane but then had a meeting changed, and that's a story?

*It is [chuckles]. Yes, barely.*

Another absurd thing was this: it was during the whole Anthony Weiner thing.[6] Basically, the House Minority Leader, Nancy Pelosi, and the Senate Majority Leader, Harry Reid, were holding a joint press conference during the scandal, and Weiner hadn't resigned yet. Everyone was thinking, 'Are we going to have an announcement about this scandal, and do we know the fate of the congressman?' Well, Pelosi gets up to the podium, and she tells the journalists (before anybody gets a chance to open their mouth): 'Guys, we had a very good productive meeting, we are here to talk about job growth, economic expansion, green energy . . . a whole host of issues facing the country, but we're not going to be answering any questions about Congressman Weiner, so don't bother asking.' Cut back to the studio, and there's the anchor looking grave, who says, 'It brings us no joy to have to cover this story, we'd like to be able to talk with you about job growth and economic expansion . . .' and he lists all the same things that she said they would talk about, but they didn't put the camera back on the press conference, they went away to do more Weiner coverage. That's insane.

*Yes, but that's true. You're absolutely right.*

Why is that? I'm not the one who coined this, obviously, but it's quite true that information has, in very many ways, become *infotainment*. It's a problem in a free society. Publius, the anonymous author of *The Federalist Papers* [it was really Hamilton, Madison and Jay who chose Publius as their pen name], constantly talks about an enlightened and energetic republic, citizenry, being the last bulwark against tyranny. The fourth estate has historically been important because we have informed and galvanized citizens. Now, we're distracting and sapping the energies of the people. I don't think that we're doing that on purpose necessarily. Certainly not as journalists, it's not what we're interested in doing. But we're doing it none the less. The consequences of that for ordered liberty and for the future of ordered liberty are things we cannot postpone thinking about indefinitely.

*We need to draw to a close. I always ask people what they see on the horizon. What do you foresee in your professional area in terms of digital technology within the next five to ten years?*

Boy. I'm not a Luddite [laughs] but I'm not the opposite of a Luddite either.

*You're not in love with the technology?*

I am in many ways a barbarian. I know how to use technology when it's user-friendly enough. In five years' time we will either have decided to really wrestle with and figure out how to use technology, or not. I am thinking of Antonio Spadaro's book, *Cybertheology*. One of the main points that he makes is that the internet, for example, is not something that we *use*. It is one of the contours and one of the features of the space of the universe that we inhabit.

*Yes, he makes that point. It's a good book.*

Technology is still something that we use; the devices that we use to get connected. I'm almost tempted to put this into a Hegelian register, such as the following: we're either going to figure out a way to reassert our mastery over technology, or we are going to end up abject slaves to it. I don't think that we have much more than five years left to begin to think consecutively, systematically, deeply, and with a view to action before it's too late. We were talking earlier about students and their inability to process more than 140 characters. They don't have a critical capacity and critical capacity is not something that can skip a generation without the next generation being incapable of anything but slavery, literally.

*I follow you.*

In many ways our whole educational approach with an emphasis on technology and on using technology is something that looks like a *preparatio servitutis*, a preparation for slavery, for servitude. Look at higher education: a period of time spent at an institution of higher learning is no longer considered valuable in and of itself. Colleges and universities are themselves concerned with giving people the skills to be the workers of the future. Their discourses about being leaders and developing leadership and excellence are just wonderful buzz words. Leaders of what?

When I was going to college, we still had the idea that it was a privileged time, that you would go on to become a better person, a fuller person, a complete person. The purpose of higher education was, in the Aristotelian sense, to actualize your humanity. I talk like that in rooms full of people who are in education and they look at me as though I'm a dinosaur or, worse, a clown. I think technology can be equally used to develop critical capacity. We need to find ways to do that. I don't think it's hopeless, but the situation is pretty dire.

*I think there's a tendency that you're identifying which is very, very true. It's inevitable, like a snowball effect, almost. And you're absolutely right. Now, I think that that trend can be reversed, and I hope it will be as the technology becomes more invisible and more widespread. There may be a certain obsession with technology which is actually decreasing.*

The newest thing is already 'old hat' and so there isn't this sense that we have to have the newest toy. We'll see how these issues evolve in the coming months.

*Thank you very much, Christopher.*

10.

# *Industrial Automated Systems*

*(Franco Stefani)*

More than forty years ago, a young engineer foresaw the rise of automated systems to make products, specifically ceramic tiles. His name was Franco Stefani, and he started a business in northern Italy which now is operating in twenty-five countries worldwide.

I had met Franco through a mutual friend who suggested that I ask Franco about his dream which has now become a reality. When I learnt that it was Franco's company that built Coca-Cola's automated warehouse at their headquarters in Atlanta, Georgia, I was convinced that I needed to talk with him. I was not disappointed, for Franco's narrative coincides exactly with the historical moment in which digital technology came into its own. Our conversation lasted almost five hours over several days. Here the reader will find a condensed version of how a young man's vision became a global success story.

* * *

*At the beginning of these interviews, I usually ask people to give us a brief biography.*

Let me tell you a bit about my roots, where I come from and how my adventure started. I was born into a peasant family; my parents crafted Parmesan cheese in the hills near Sassuolo, in a tiny village called Varana. At the age of twelve, I left home to come to Sassuolo, where I started studying in a vocational school.

*At twelve years old, were you alone or with your family?*

At first I was alone and I was hosted at my parents' friends, then six months later, my family joined me.

*At the time, I guess a twelve-year-old boy was quite mature.*

Yes, indeed. I can tell you that I have always led a very independent life, because I had parents who were very busy in the production of that famous Italian cheese. It is a process that involves working almost fourteen hours a day. Especially in those years, it was all manual labour. I had to get by on so many occasions. My father and mother came to Sassuolo and decided to open a dairy. I helped them selling milk in the afternoon by transporting more than ten gallons on my bicycle. I was a sort of 'milkman', as one used to be in those years. There were no supermarkets. Regarding my education, I did three more years of professional training at the vocational school and then at the age of sixteen I started working at a local firm that produced tiles, called Ceramica Marazzi. I was hired as an electrician. It was a great opportunity for me as I had a strong passion for all kinds of electronics. I also enjoyed playing around with amateur radios and model airplanes. So, even though I was only an adolescent, my curiosity and attitude led me to the creation of the very first transistors and electronic valves. At the company Marazzi, I even applied the concept of electronics to manual screen-printing lines.

I installed a small piece of automatic equipment on these lines, at the age of seventeen. At that time, my department head understood that I was passionate about automation and gave me the opportunity to try out this application. It was a great success, as with that application I increased productivity on that line by almost 50 per cent. They awarded me three months of extra salary. I used timers that gave a specific timeframe to those who did the manual operations. If the worker did not complete the action in time, a red light would switch on. Finally, with a little automation we boosted the performance of the line and eased the task of the operators. At that point, I realized that my key goal was to transform industrial processes into automatic processes. It was my pioneering attitude that led me to the creation of other small automations for the Ceramica

Marazzi. Then I designed a machine to automatically run the silk screen printing system and thanks to my application it became fully electronically automated. At the age of twenty, I realized that this was going to be the future of production. The experience in the local firm opened up a new scenario and made me understand that it was time for me to establish my own company.

*And that was when you were twenty years old?*

Twenty years old was considered old enough at that time to create your own business. I went to my father and told him (in our local dialect), 'I've decided to go on my own. I want to transform my dreams into reality and obtain results. I'm ready.' I founded my first company in Sassuolo which was called COEMSS (Electromechanical Constructions Stefani Sassuolo). In the following years I founded SYSTEM (Stefani Electronic Mechanical Systems). I developed the most sophisticated machines for rotary screen printing and sorting lines, by applying electronic devices. I also had the chance to hire some friends of mine. Ever since those early years my attention has been focused on Research and Development. My idea was to create one of the most innovative companies in the world. And I did it.

Continuing on this path, everything that I was earning from the activity of the company I would put back into the search for innovative applications and high-end technologies. I have never pulled out even a penny for myself [all profits go back into the company]. During the same period, I started dating my future wife. At age twenty-five I married her. We often talk about 'multi-tasking'; well, I was always 'multi-tasking'. I would have to complete four or five tasks simultaneously. I have continued to carry out this mission, and have been fortunate to have been always supported by my passion.

*Does your company operate only in Italy, or around the world?*

We are all over the world. Our customers are everywhere in the world, we have no frontiers or boundaries. Eighty-four per cent of all our business consists of exporting our machines and plants.

*In how many nations is System Group now present?*

We are in twenty-five countries worldwide. System is an international group that offers its solutions to a variety of sectors. We are just about everywhere.

*Impressive. Before, you were talking about the link between mechanics and electronics: can you explain a bit more about what you mean?*

Before, machines were – from the kinetic point of view – rigid systems: there was the driving motor and then there was the whole kinematic system to the process. With the advent of electronics, controlling devices, processors and information technology, we have been able to make autonomous, related entities, mixed and arranged through an electronic process. In other words, the electronics become 'supervisors' of many small movements, and of many intricate processes. This evolution is now called 'mechatronics'. This means mechanics is at the service of electronics, and electronics guide the mechanics involved and together they constitute a machine or a process organized in a precise way, through a succession of analyses.

*Although I realize that this is complicated, could you give me an example of some machine that you are particularly proud of?*

We are specialists in machines for printing and decorating ceramics in digital form. These machines begin from a knowledge of digital technology, that is, as processing images in the digital domain: pixels over pixels. We then proceed to the storage of the image, first by recognizing and then by scanning. We can apply this, for example, on a natural stone: I scan this rock, I transform it into digital information, I place it in the memory and from the memory I elaborate the image. When I have the digital file, I compile it over an elaborate memory, a memory base, with a processor that manages in real time the inkjet printers, firing the different colours on to ceramics.

*Looking back, could you tell us about your experience in Palo Alto, California, in the 1970s?*

I was always fascinated by electronics: a universe that makes great use of engineering, physics, mathematics, software and hardware. Electronics today can be found in almost all industrial fields. From the very beginning of the semiconductor revolution, I have always found a passion for its development. In the summer of 1978, the year I started to build my company's factory, I went to San Francisco in order to visit an international trade show. An engineer of the company came with me to the show in the United States. In the 1970s everyone was speaking about the American dream and about California, mentioning Silicon Valley, and as I was so curious I wanted to experience it first-hand and see what Silicon Valley was all about. Arriving in San Francisco, we rented a car, and started exploring the city. I have always been fascinated by aeroplanes, too. Under the famous Golden Gate Bridge, there was a seaplane base with aircraft to go on a city tour. In those days, with one hundred dollars you could get an hour-long flight. I said, 'Let's go see the city by seaplane.' We took the seaplane, and we toured all of San Francisco, going above the pyramid, next to the skyscrapers, in and out of the huge buildings there. It was amazing: a beautiful and unique experience!

*Back to the Trade Show, what happened next?*

We went to this show which was one of the most important venues of those years. The big brands were represented there: Texas Instruments, Motorola and Intel, just to name a few. We began to make the very first contacts with Texas Instruments, and we brought back to Italy the very first sixteen-bit processors. Then we started to understand what electronics was all about and we began building the first boards with CPUs inside.

*Precisely the very beginnings. Impressive.*

I have experienced an historical change. Even the semiconductor manufacturers did not know what these objects were for and why they were making these processors. Only later would they find out. In fact, many applications came only afterwards: they were placed inside phones (still analogue) and in telecommunications. They were made out of basic processors, which were open with internal memories. Slowly, the use of the processor took shape, tailored according to the most diverse activities.

Now, machines can do the hard work, but they cannot replace the human being in the willpower to do something.

*Tell me a little more of what you mean by 'the willpower to do something'.*

The willpower to do something is imagination. The feeling that drives you into action. The concept is that no one can do what you are thinking, exactly as you are thinking it.

*What inference can be drawn?*

Based on this reasoning, machines are created to be at man's service. I believe in the training and education especially of young professionals. This aspect is fundamental to develop new applications and technologies. Man will never stop thinking and machines can give you instructions and directions. At the end you have always man's will.

*Can you give me an emblematic example?*

Well, I did have an experience this summer during a dinner at the home of friends with their children and grandchildren. We were at a beautifully set table and the kids were at another table. After dinner, the children sat on the couch and each of them had their own device such as a smartphone or tablet, and small games suitable for their age. Each of them became comfortable in their place and started to play: the room fell silent. I saw them all sitting, all quiet, each with his or her own game. They had entered into a fantasy world, happy that we adults did not matter any more. They were completely isolated with their gadgets, built by great engineers for both kids and adults.

*Yes, through the gadgets?*

Yes, through information technology. I'm enthusiastic about how technology can be so intuitive for the young generation. It's a process that I'm still analysing and I like to observe it in details. This change can have advantages and disadvantages, it all depends on how it is managed. We cannot forget the real world we live in. Technology is a precious ally and can be of great advantage. I'm a fan of the technology.

*At the same time, however, you are cautious about some trends.*

Yes, because I think the human aspect is very important. For me, people are extremely valuable. In my work, for example, I give much importance to my collaborators. I feel responsible for all these human resources. More than 1,500 families depend on my firm; that number climbs to 2,500 if we can consider those touched indirectly by System.

*Does System Group have difficulties managing cyber security?*

We really don't have much risk with cyber security because our projects are fairly protected. Obviously, we have had several intrusions in the past, but we have strong filters and firewalls and we are well updated in terms of security. Our work must be protected, even though our business is aimed towards a very narrow set of clients. On the other hand, for a business aimed at a much larger clientele, they would be more vulnerable and would become easily targeted. Our product is very specialized, and it is difficult to place on just any market.

*And speaking of what's to come?*

We are anticipating many changes: we must embrace them. There will always be new proposals around us, which may become standardized because of their convenience – it is always man who determines the standardization of processes and of products. If these changes are beneficial to man – in terms of comfort, or of necessity – man works them out in terms of their cost/benefit analysis.

*Still on the subject of technology. Have you seen the app that connects to the iPhone 6 in order to measure different parameters of the body?*

Yes, it does so by means of sensors, which Bluetooth then transmits as data to the phone.

*Do you foresee the day in which a gadget or a microchip will be implanted into the brain, and will interact directly with it?*

Right now, it is still science fiction: if one looks around, reads a little or watches certain films, you could say that something is already here. But I would say that it is not accessible yet.

*Could this become a reality in twenty years, or in thirty years?*

From a technological point of view it can happen, because even now we are analysing some semiconductors – in a state that we call 'organic layers' – which are compatible with organic material and can be powered by the body. There are many steps that still have to be completed.

*So, for me that means that we will have this technology soon.*

It means that we will have certain sensors which give us information, and which connect to databases to provide us with useful indications.

*The indications?*

Guidance on how to live better. For example, a device such as a cell phone, which everyone already has, can give you health advice also.

*If you could give us a brief overview of your company's philosophy, that would be enlightening. It seems simple, but in reality it is somewhat complex.*

My business philosophy is based on knowledge. The knowledge that one possesses is the wealth of knowledge that one has synthesized and that must be expressed, so that later one can place a product on the market. This requires capital goods (my company has been working on this type of goods). It is necessary to first of all know the state of things and then improve the technological part. One must analyse the contribution that can be made by introducing a product or process (something that changes things, something that is innovative), leading to the creation of machines or automatic systems that can create new markets. If you manage to do this, you create new products, you create the market and attract the customer's attention. The human heritage is fundamental in the creation and fulfilment of ideas. That's why here at System you will find electrical engineers, mechanical experts, physicists and chemists. Collaboration among all these sectors will allow you to create high-quality products and high-level jobs. In other words, man is at the foreground of the activity, he is the key; without people, all the buildings you see around you would be useless. We believe in the centrality of innovation and the differentiation of products is a cornerstone for our manufacturing activity.

*You must have some very competent collaborators, right?*

I have a number of collaborators, who above all believe in me; they believe in the projects I am contemplating and together we think about where we want to go. After that, each one then develops his or her own specific field. I provide the guidelines, I am personally interested, I want to understand what must be done and how. That is why I am helped by a specialist: I want to understand at least the essential features.

*Here at System, if you walk around and speak to the employees, you will discover that everyone is excited to work here.*

People are enthusiastic about working in System as they feel they are part of a unique world. They are also free to act as we are using a method based on responsibility: everybody does their duty, completes the task and no one controls you, except for the system (in general). We have an information system that wants to know what you are doing. All this brings a great sense of self-accomplishment and shared ideals.

*One of the most difficult aspects that some have told me about refers to working in many countries. Given that you are present in many countries, how do you address these difficulties?*

Yes, indeed, difficulties can arise. In those countries where we have other locations, in addition to an Italian executive who resides in the country, we select local people who know the local culture and who help in deciding strategies and ways of operating. We create a structure that takes into account the local environment. It is necessary to speak the same language and know the customs and traditions.

*And in China, for example?*

In China, there is an Italian manager for us who went there with his family and manages the business with one hundred employees and runs it very well; in addition, there are Chinese directors who work with him.

*What you describe seems like a perfect combination of what is called 'glocal', i.e., global and local.*

When you trade with other nations, you have to think in a different way. You need to be more open in accepting the conditions of the host country.

*I was just reflecting on the value of people in a company like yours. You might think that the machines are the priorities, investing in machines and doing research. Instead, what I perceive is that you invest more in people who then will do this research.*

People are the priority. System has always been characterized by its commitment to human heritage. A pioneering approach must go hand in hand with the training of professionals and valuable collaborators.

Let me tell you about a wonderful experience that is happening here right now. I discovered a young talent from a technical school, and I followed him for three or four years. When he turned twenty years old, I hired him here at System. I started having him do specialized courses, and I began to entrust him with some tasks concerning a very important project. When he was just twenty-one years old, he and I filed for our first patent together.

*Unbelievable! Where is this guy from? Is he a foreigner?*

Yes. He is from Morocco. He possesses a unique intellectual brilliance. This guy is carrying out a project which happens to be the most important project that System has ever launched. Every day he brings new ideas to the project, constantly carrying around his laptop. He works ten hours a day with me in the factory and does marvellous 3D designs. He is capable of operating very sophisticated machines that utilize special lasers for micro-welding. He is also taking some specialized courses because we are developing some very sophisticated technology.

*He certainly sounds like a very special person.*

This is the value, the irreplaceable value, of human beings. He is not content with this initial project, but now wants to go further and launch new initiatives. All of this has been built with hard work and commitment. We must never forget that.

*Impressive! We tend to forget that even in a company based on machines, there are always people behind them who are responsible.*

Yes, there are always people.

*We cannot hide from the fact that many people are afraid of machines. Recently, the national press in Italy covered an Oxford University study which stated that in twenty years 35 per cent of jobs will be taken over by machines. They went on to make a list of the most vulnerable sectors. We get more predictions on this issue almost every week. Do you think such predictions are unreliable or are they something to reflect seriously about?*

No, those predictions are not unreasonable at all. Just look how we have advanced since the 1950s and how many things have changed dramatically. Recall the fact that, years ago, one person could produce only five square metres of tile a day, expending a tremendous amount of time and energy, and now machines produce 500 square metres of tile much more quickly and more cheaply. Times change, but the hand and the head that moves the machine is fundamental. In System skills are the key aspect.

*Would it be too bold to say that the new channels of communication today are digital?*

We have antennas all over the place, those are the channels of communication, without which there would be no chance for us to communicate or achieve results. We are now in a digital world and this is the driving force. Changes are occurring all the time.

*Every change is scary, I think, especially if you're used to the old way of doing things.*

I would never want to express fear about the progress of technology. We welcome technology, we deal with it and we try to understand it. One has to manage technology effectively, accepting it and using it to the fullest. This is my way of thinking.

*One final question that I often ask: can Franco Stefani be replaced by a machine?*

No, this cannot happen, machines always need an observer, and that is the role of the humans. If there are no people, then machines cannot exist. Man has to be 'ruler' of what happens around him. Machines can help us, just like all the tools that we use. For example, who makes cars travel in one direction or another? People do. Machines have certainly solved many problems for humans, they have helped man to create things ever more precise, ever more useful and technologically sophisticated. This is all part of the construction of human thought applied to machines.

## 11.
# *Existential Risk*
## *(Anders Sandberg)*

The very name of the place is provocative: the Future of Humanity Institute at Oxford University. Who would not be intrigued? We all need to give the future of humanity serious reflection. When I wrote Anders an email asking him to talk with me, he responded immediately and we had a delightful conversation in Oxford.

Anders Sandberg is one of the directors of the Institute, specializing in technology and existential risk, as well as considering possible futures for humankind. Although he travels a lot, he spends most of his time at the Institute researching and, above all, thinking. He is originally from Sweden, and can be considered part of the 'Swedish Gang', which also includes Nick Bostrom (quoted several times in this book and who is president of the same Institute) and Max Tegmark from the Future of Life Institute in Cambridge, Massachusetts. Anders has a wonderful sense of humour: he was even capable of speaking about nuclear wars and climate catastrophes without succumbing either to cynicism or to hyperbole.

* * *

*What are you currently working on?*

My main project right now deals with systemic risk-modelling in reinsurance.

*In reinsurance?*

Yes. It is like making a market for acts of God. If an insurance company insures your house, they have to pay you if it burns down or gets flooded. Okay, that's no problem, they insure a lot of houses. They average it out and they make nice money. However, an occasional hurricane comes along and a lot of houses get flooded. That's bad news for an insurance company. So, of course *they* have insurance. They go to the reinsurance company and say, 'Hey, Miami got flooded and we owe a lot of people a lot of money. Could we please have some of our reinsurance money?' and the reinsurance company grumbles and gives them a billion dollars or whatever it is on the contract.

*Who is the reinsurance company?*

They are mostly companies you rarely hear about like Munich Re or Hanover Re. The one I'm working with – Amlin – is kind of a mid-sized British one. Again, you rarely hear about them because, as a consumer, you only deal with your insurance companies. But these are the reinsurance companies. That's where the real money is, because they need to have more money than the insurance company. They are the ones that have the big skyscrapers in the centre of London. Let's say you are a reinsurer. You're sitting on liabilities for tens or hundreds of billions of dollars. If a hurricane hits, you need a lot of money rather quickly, what do you do? You want reinsurance. They're also reinsuring each other, and if you deal with it right, everything is really fine. If you do it wrong, well, there is another book in that pile which is about how the Lloyds market almost totally imploded in the early 1990s because we got into a kind of circular arrangement. Everybody owed each other more money than they could pay. It was very similar to the financial crisis, but they resolved it locally, and nobody outside ever learnt the lesson.

*I never heard about that. But it sounds like we are deceiving each other: I think insurance itself is a risk. It's a calculated risk.*

Yes. If your house burns down, it's very bad for you. It's actually smart to pay a bit of money every month, because that certain loss of funds is way better than the much bigger gamble of total destruction. On the other hand, for insurance companies, since they have a lot of houses, it averages out. They can do statistics. You hope this is where our project comes in. If you do have everything right, and everything works as proposed, everybody should be happy.

Look at life insurance. When it first arrived, people felt outraged, and now people have forgotten how creepy life insurance actually is. Except, of course, the insurance people being naturally gloomy, they look out at the bank skyscrapers next door, and say: 'Hey, those guys are just as smart as we are. They fouled it up badly.' Okay, what's the problem here? 'Well, we're using mathematical models which are not true.' Everybody initially loves to point that out; of course our models are approximate, then you need to use your own judgement, and then they get worried. 'We're all using the same models.' Everybody's using the same hurricane models and so you arrive at the probability of Miami getting flooded. 'Okay, and if that model's wrong, which we have good reason to believe, we're all wrong in the same direction. Uh-oh.'

*That's where the big money losses come in.*

Yes. That's what they're scared of. This project is about helping them think about how they think about models. It's very *meta*.

*Very meta. Now, can we just go back a little bit to ask, what is the Future of Humanity Institute?*

We were founded almost ten years ago. We're soon getting to our anniversary. Nick Bostrom was working as a philosopher here.

*In Oxford?*

Yes. Originally, he was at Yale University, where he produced a number of interesting publications, as you probably know; for example, about the move for human enhancement and questions like 'Are we in a big computer simulation?' and other technological matters. Then James Martin showed up. James Martin had made his fortune in the 1970s and 1980s in the computing business, and he was one of the old-fashioned engineers who invented a lot of computer engineering. Once you're fabulously rich, and you have you own island in Bermuda, what do you do? Well, maybe you get a white cat and try to kill James Bond. You could try to turn into a superhero, or you try to save the world, or you're really bored, you just try to become richer and play more golf. But James Martin decided, 'Let's donate a lot of money to Oxford to set up an interdisciplinary school to solve all the big problems for the next century.'

*This is due to James Martin?*

Yes.

*And he attracted Nick?*

Yes. James Martin came to Oxford and started setting this up. Then, a fellow and a colleague here from the philosophy department, Professor Julian Savulescu, began helping out. He has had some rather interesting arguments with Catholic bioethicists.

*I've heard of that.*

This building here is the kind of ground zero of consequentialist ethics. Julian told Nick, 'You might want to meet this guy,' and James and Nick hit it off really well, because Nick is interested in the long term and in the big picture about the questions of humanity. We try to solve the big problems facing humanity, so we have institutes concerning climate change and schools dealing with nanotechnology and stem cells, ethical thought concerning armed conflict. Having this kind of long-term perspective is very important. You want somebody sitting at the top of the tower looking at the horizon, trying to see if there's something else coming out.

*That's what Nick's philosophical background trained him to do. But he's also very, very well read in terms of the technology and the science.*

Oh, yes.

*Some of his recent book is a little difficult to read. But he acknowledges that at the beginning. You see there are some parts that are going to be very techno, you can't get through. And I tried reading two or three times a couple of passages and I just gave up. But I was more interested in the third and fourth chapters, I believe, where he talks about existential risk.*

When you start thinking big, you start to wonder, what are the really important values? What are the things we actually should care about? And existential risk is one of the more obvious ones.

*Of course, especially if, like he says, superintelligence could be worse than an atomic war.*

Exactly. Around here we have a saying: 'Yeah, it's just an atomic war.' It's bad, but it might not be the end of the world. Actually, I've been having a long-running argument with Andrew Snyder-Beattie; we're co-authoring a paper about the probability of nuclear war. We're arguing about who's the optimist here.

*Well, if I can interrupt briefly . . . a month ago, I read that the Doomsday Clock, which is sponsored by the International Atomic Energy Agency, is now one minute closer to midnight.*

Yes. I think they are roughly right that, yes, the world has become much scarier over the past few years. And that's worrisome. The problem is, of course, how do you arrive at the actual probability? There are various methods of calculating that. You can build up scenarios, you can make possibility trees and try to assign probabilities. You can be clever and realize, 'If I say 4 to 7 per cent, I'm actually uncertain, I should have probability distributions.' You can build very, very sophisticated arguments, but in the end, of course, it is all going to be very uncertain. If the Cuban missile crisis had ended in nuclear war, we wouldn't be sitting here talking about this paper. So, the probability testament is a bias.

*But there was a high probability.*

Kennedy said it was 50/50. And afterwards, some evidence has come forward showing us that, yes, people were overconfident about how safe they were. There were actually quite a lot of scary things happening at that time. Sometimes, we've been saved by chance.

*It's so amazing. Chance. It was just luck.*

We don't want the fate of the world to hang on blind luck. At least we want to reduce the amount of existential risk. You can't avoid all of it. There is a fine probability of everybody dying of natural causes. It's very low, you can calculate it and you cannot get rid of it. But on the other hand, you have stuff like nuclear war. Adding the red telephone, that helped a lot. Disarmament helped. Nuclear proliferation is a rather bad thing, because now the game theory that kept things stable gets fouled up. Our paper is trying to look at the range of bias: I am arguing that this is actually a pretty big effect and the world is really scary.

*What is?*

The fact that our existence is kind of biased in terms of the probabilities, because we would only exist in a world where there hadn't been a nuclear war. Our evidence of seventy years of no nuclear war shouldn't make us very calm at all. Even these near-misses, we should actually interpret them as much scarier than they were. Andrew thinks the actual probability effect is small, because he thinks nuclear war can be survived. The real threat is the nuclear winter, and we don't understand nuclear winters very well. The scary part here is that there was a lot of discussion about this. Those discussions contributed to Reagan actually going for disarmament. Yet, later, people dealing with the first Kuwait war became convinced that they weren't such a big problem. The cooling of the ground in the desert was really serious, but it never got up into the stratosphere. It actually looked pretty nice. Then, more recent climate simulations showed us, no, actually it takes a bit of extra force to get it up there. But once it gets up there, you get a very, very bad scenario. Still, you might argue (like some of my colleagues do) that there's still going to be some people holed up in the Walmart logistics centre, with a lot of canned food. It's kind of no guaranteed extinction.

*A nuclear physicist told me that at White Plains [where they tested the first nuclear weapon] there is still residual radioactivity. So much so that they fenced it off. You can't get in. And this happened in 1945.*

Ground bursts are kind of nasty, because all the atoms do get a lot of extra neutrons. You get isotopes that remain for a very long time.

*And this was a 'small' bomb.*

Yes.

*Trinity, I think, was the name of the test bomb.*

Yes. They even named a mineral after it: trinitite.

*Oh, really?*

That's a kind of a green glass you get when the desert sand gets fused by the heat. It's actually regarded as one of a few man-made minerals. I know of one other, that's a weird conglomerate of frozen whale blubber and sand that you find in some parts of South Georgia.

*How did you come into the Institute here?*

When this began, we got financing to start the Institute, but of course we needed to have projects and project funding. There was a big European Union project about human enhancement and, around here, we were doing the cognitive enhancement ethics part. I applied for a job. I have a background in computational neuroscience. I'm interested in how to improve brains, and what the social and ethical implications are of that. Since I also knew how to make web pages, I was web managing for the project. And I decided, well, why not try it? Since then I've stayed here, I've managed to get other project funding and I've been kind of keeping active in the Institute.

*And how many people are you here?*

We are about fifteen people. I think we are going to expand a bit now because we've got a European Union grant to look at some questions concerning the precautionary principle and stuff like that.

*Now, correct me if I'm wrong, but I believe I read that Elon Musk gave Nick $10 million.*

That's not quite right. Early this year [2015], there was this very interesting conference in Puerto Rico. Many of us were there.[1]

*Did Elon go to Puerto Rico?*

Yes. On the final day, he said, 'I'm very convinced that this is a relevant problem. I'm going to give $10 million to do research in this field,' and this has been managed by the Future of Life Institute, an outfit over in Boston run by Max Tegmark, the cosmologist.

*Because he was the one that organized the meeting in Puerto Rico?*

Yes, yet another member of the Swedish conspiracy (!).

*Swedes. But I think he's very close friends with Nick, is he not?*

Yeah, he is. Nick and Max had a significant paper in *Nature* a few years back about the doomsday probability. It's one of the funniest ways of estimating a risk I ever encountered, because it uses the age of the earth and solar system to estimate risk, which seems totally backwards. It's a very odd kind of argument, but it works. The Future of Life Institute is actually today going to announce where this research money is going.[2] I have been posting recently also as a response to the hype around artificial intelligence. What should we actually do about all the hype? Whose fault is it? And, unfortunately, the nasty problem is it's everybody's fault. It's not like we could say, 'Yes. If only those researchers could shut up. If only those journalists weren't hyping.' They want to make a story that people go and watch. And meanwhile, if you have a start-up in Silicon Valley, of course you want to claim that you can cure cancer, and make artificial intelligence, and clean the atmosphere. You're going to claim whatever it takes to get money.

If you're Google, well, you're going to try to maintain a nice, good image too, and then your competitor Baidu will claim, 'Oh, we can do better than Google. Ha, ha.' Meanwhile, of course, there are some serious academics that never speak up, and they have a hard time getting funding. If you speak up too much, then you might not get funded, because now it's obvious that you've gone too far. But if you just 'one up' someone, it might work. The end result is that everybody tends to amplify each other. Then you get people responding online, blogging about something cool or forwarding a link. We have this epistemic system in our civilization that is broken. It's at least not working very well. We need to improve it. I think this is actually one of the biggest civilization challenges we face today. Here, digital media are quite essential, because we're very much part of both the problem and the potential solutions.

*I think Google is looking at this. Now, could we go back again to Puerto Rico?*

Yes. Sorry.

*There were about, what, sixty people there? What was the environment like?*

Basically, they took a resort hotel and filled it with both artificial intelligence engineers, many corporate members and academics who research artificial intelligence, or research risks with artificial intelligence. There were some activist organizations, like people who are against drones, and robots that kill. There were economists who were interested in the questions of employment and automation.

*That is huge.*

It is. Another fellow there (although he's not at FHI any more, because he's got his own little e-research project now) was Carl Frey. He and Michael Osborne wrote a paper in which they predicted that 47 per cent of all jobs are at risk because of AI.

*That has been heavily quoted, it's a great paper.*[3]

Yes, it's a great paper. You can quibble about the methodology, but I think they're roughly right. There are a lot of skills out there that can be automated. That's not necessarily their entire job. There are components of different jobs which are hard to automate, but the numbers and statistics you can, of course, throw to a computer. The 47 per cent is not as exact as one would like it to be, but it gives a sense that a lot of jobs will at least be changed profoundly. I think it's very hard for some groups to understand, especially many politicians over in Sweden; our government simply doesn't get it.

*I was looking at the website for the Puerto Rico Convention, and some of the most important people in technology and philosophy of technology, like Nick and yourself, were there, but I saw no religious figures.*

Yes, that is true. I think part of the reason for this was that it was framed more as a technology risk issue, which is interesting of course. Again, from your perspective it's totally obvious, but in that case we definitely should have a priest involved somewhere. I think this might be a little bit like the problem they had with the Asilomar Conference. Do you know about that?

*No, I'm not familiar with it.*

It was the famous Asilomar meeting in 1972.[4] This was the classic debate about science trying to do the right thing, and self-regulated, potentially dangerous, technology. The standard story is, of course, that scientists realized the importance of genetic engineering, they called together this meeting, people discussed it, they hashed out a protocol, and they managed to reach a consensus on implementing it. Overall, it has worked pretty well.

*It hasn't been obligatory though, has it?*

No, it was very much voluntary. Everybody who was involved agreed on it, and it was a good idea. Looking back, they did the right thing. Except for one thing – they didn't have any ethicists around, and had they invited anybody religious, it would have been even more bizarre. Some people actually said, 'Well, we need to agree on key issues, but we will not agree on these issues if ethicists are around.' Which might indeed be true. To some degree, you would have a long ethical debate, and hashing out what should and shouldn't be done: that's not going to happen over a weekend. You can kind of agree on what safety protocol to use in a lab. But you won't be able to answer the question, 'What rights do we have here?' This is, of course, deeply annoying, because to some extent you need to have all possible voices involved. But you also need to have something that's workable enough to get something started. Therefore, I think, in the case of the Puerto Rico meeting, we didn't agree on some big protocol there, but it's mostly a statement of intent. It points in a rough direction that may make sense.

*But that was just the first meeting.*

Now hopefully we're going to see this Elon Musk-funded project coming up, and the aim is very much to get people involved, especially people in computer science who normally would never touch ethics or anything else like that with a ten-foot pole. I vividly remember a few years back when a Google engineer, who was working here, asked: 'Is there an ontologist in the house?' Gina poked her head out of her office and said, 'Certainly,' and then there was a very enjoyable half-hour when she explained Kantian ethics to poor Daniel and Daniel just shook his head. 'That's not computable. We can't possibly put that into a program.'

*That's exactly the point.*

Yes. Kant would have smiled and said, 'Yes, sure,' but Daniel would have said, 'That means we cannot make a machine that is going to behave itself.' It might not be a moral agent, but you certainly don't want it to be dangerous. So, maybe you can do something scaled down, but of course Kant would say, 'Yes, it's not a moral agent,' and they would probably have left it at that. Here, we are constantly pointing out that something might not be a moral agent but it can still be terribly dangerous. A car is not a moral agent, but you don't want to get run over. Philosophers quite often focused on the moral agents, but that's about it. Maybe we can have a fun conversation about when robots should have rights? But the problem is, we have these weird systems which are not moral agents. We create them. We cannot control them fully, but we want them to be safe, and a lot of the problems are very similar to morality although they're not exactly morality proper.

*That's a can of worms. Did you discuss these things in Puerto Rico?*

Yes. There were many discussions about these issues. Some were on a very high level and abstract, others were much more concrete. Could you have self-regulation in the industry to have rules put into machines so that they don't do nasty things? Is it even possible to make rules like that?

*It is. You can give the machine the possibility (which DeepMind is doing) to change its goals, to learn and therefore to alter its own program.*

Yes. This is one of the most obvious problems with the Asimov-style laws. You give a robot laws and it's going to follow them and later change them. But if those laws are wrong, now it's going to do something very stupid, and in reality, you want it to learn because it needs to understand that, 'Oh, that's not what my creator intended. I want to do what was intended, rather than what he explicitly said.'

*Do you think a machine will ever be able to make that distinction?*

You can definitely do that in a very simple form. Now, I expressed it in a way that sounds like thinking philosophically, but that's mostly because it makes it easy to explain. I think in practice what you would have is software code evaluating different options, including evaluating the option of modifying itself. There exist models of software that actually do this. Self-modifying software has actually been around since the 1960s except very few people use it because most of the time it is not a good solution.

*It doesn't get it right?*

Well, in most cases, if you're not intelligent and you modify yourself, you're going to do something stupid.

*As an example: the human race (!).*

Yeah, exactly. The problem is also that, in most cases, self-modifying code is more like this: Here's an instruction to jump over there. But I can modify 'from here' to mean 'add those numbers'. Now, the program changes, and when the execution comes over there, it does something different. This is horrendously hard to debug. It really messes itself up in weird ways. But you could meet with Professor Jürgen Schmidhuber, who lives in Lugano, Switzerland. He's written a few papers about what he calls a Gödel machine. This is a piece of software that proves theorems about its own code. And if it can prove that 'If I change myself in this way, I'm going to be better', then it will do it.

*It will do it.*

Yes. Now, it is nice theoretically. I don't think it can actually do anything useful on a real computer in a real amount of time, but it's nice mathematics. It's kind of a proof of concept. Speaking about reinforcement learning systems, you now have something interesting. You have something that cares about maximizing a reward. That's essentially the only thing it cares about. That's how it's structured. But what gives it a reward? The standard method is when points are awarded in a computer game. That's a very easy and direct thing. But there are models where you can actually learn other things to be rewarding. It's a bit like how we grow up. When kids do maths problems, that's not rewarding. But then we get the pat on the head from an adult saying, 'That's really smart. Good work.' And we learn: 'Hmm, it's kind of rewarding to be good at maths.' Over time it becomes something that can be run on a computer also. My brother was talking maths at the breakfast table because our parents were very much humanists and didn't understand weird stuff like maths. We were happily talking mathematics because we found it enjoyable. We had not created an internal reward for doing this. It is very much like Aristotelian ideas about how our habits can create virtues. With them, they allow us to learn higher-order things which might previously have not been possible. We actually construct higher-order rewards.

*Do you think we could induce the machine to act like a moral agent (although I agree completely with you, it's not a moral agent), by giving it a kind of meta-ethics? For example, what if we were to say, 'Maximize the greatest good for the largest amount of people'? Which is not very human, because we're very interest-orientated. We want to maximize our own interests or those of our family, or community, or corporation.*

The problem with that meta-rule is, of course, how widely it can be applied. There is an infinite amount of possible things it could be doing. Thus, it needs to allocate its effort somewhere. That's going to give it a kind of personality or style in how it's exerting its efforts. Let's think about a very small problem, an autonomous car, for example. It's driving along and a child runs out into the road. It should swerve, of course, to avoid the child. But if there are one hundred people around, now it's going to hit them. What should it do? If you have put in that idea of 'Minimize the number of harmed people', it might swerve or decide to hit the child. If a human driver had done this, we would say, 'You are morally responsible for that act.' We might even ask: 'What were you thinking? Why did you do that?' However, the machine in this case doesn't choose an ethical system. It just has one that is already pre-packaged. We might still argue, 'Hmm, maybe its sensors are a bit bad, it actually didn't notice that there was an opening over there it could have used to avoid crashing into people. That's a technical problem, we can't blame it for that.' Maybe if we want to blame somebody we might say, 'Yeah, the engineers who wrote that code are to blame.'

*Exactly.*

Really, they should have put Kantian ethics into it, because that's the proper way.

*Maybe it would depend on the lawsuits. There will be lawsuits.*

Of course. I actually talked to some engineers and they said, 'Yeah, sooner or later we will end up in court.' This is an interesting problem because you will have powerful AI and with rules put into it, such as ethical systems, or the meta-ethical assumptions that it will grow for itself, because it might figure out new ethical things in line with its original programming. There is an awful lot of power concentrated in something here. Assuming it's a single superintelligence which eventually gets shared by everybody, we all end up with the same ethics based on what somebody wrote. That's a problem, because now we have little room for actually fixing it, if something goes wrong.

*But aren't you trying to solve this problem here at the Institute?*

Not exactly. The basic problem with safety in Artificial Intelligence is, first of all, figuring out what the problems actually are. You cannot say, 'If we just knew the right ethical system, we could easily put it into the computer.'

*Exactly. But that would seem to be easy.*

No, it wouldn't. Suppose that we actually knew that Kant was right, that Kantian ethics was the true ethical system. We still can't program it, because some of its properties require the computer to do infinite estimations across all possible worlds. That doesn't work.

*Even your computer engineer says, 'That's not computational. It's not computational.'*

Maybe we can stay with a limited version. Now I have deliberately introduced something that's not the right ethical system. If this is a very powerful system that thinks deeply, some of the consequences are going to be beyond what we can see. Some of the consequences based on our approximations are going to be very weird. That is a deep problem, and this was in a nice situation where we actually knew which were the right ethics. In reality, we don't really know. Of course you might say, 'Maybe a superintelligence will be able to figure it out with relative ease.'

*Maybe.*

Maybe. It depends a lot on meta-ethical assumptions. Again, maybe it's just that humans can't figure out the right *human* ethics but maybe the machine can actually do it. But in that case, we need to make a superintelligence capable of ethical thought, even though we don't know which one. This is a huge risk, like saying, 'Let's make a superintelligence and hope it figures out what's right.' Because you can't have superintelligences that are practically powerful and do not follow what's right. My standard example is the 'paper clip machine' example. It has a utility function it maximizes, and it will make paper clips until it runs out of material resources. This can be very dangerous, because it could theoretically rob all of the earth's resources (which humans need to survive). It will do whatever increases that. It is programmed thus: 'Make paper clips, unless it's immoral.' You need to get that morality part in, and exactly how that's supposed to be is a good research topic.

It also turns out that you can start thinking, 'Well, maybe it should change what it believes in.' If I work on some project, sometimes I realize, 'Maybe this is wrong. Maybe this is an immoral project, or meaningless, or maybe I could do something better with my life.' I can actually change my values. That's of course core to what it means to be a moral agent. You can certainly imagine machines that can modify utilities. But this is something we've been analysing at some length, and it turns out that most of the simple models will not do that. They have a goal and they cannot change it. Since that goal is the supreme reality for the machine, it is not going to change it. As humans, we have multiple goals: sometimes they are in a total mess, they contradict each other, and that is problematic from an existential standpoint. However, it's very, very useful in terms of being able to shift goals.

*Why do you think humans are almost inherently against machines which interact in their environment?*

It depends on one's culture. In the West, our view of robots is innately sceptical, partially I think for religious reasons. 'God created man in his image', so in whose image are robots created? That's kind of weird. In Japan, on the other hand, everything has a soul. There are spirits in everything, coffee cups and rocks and air. Of course, they think, if we make a robot, it will have a spirit. That creates a different approach.

*That's the Japanese mentality.*

You can probably come up with a lot of other mentalities too. It would be amazingly fun to see African robotics, if that makes sense. There probably are some of those, but we haven't heard much about them.

*Yeah, I don't see them developing too quickly.*

There are some fascinating developments in Africa regarding social realities, because they're bringing in connectivity to the people who didn't have it: they can negotiate prices, they can figure out if they're getting fooled by the trader and so on. I think we are seeing just the beginning. People are doing really innovative stuff with banking. Even if you don't have cell phone coverage, you can use a camera in a cell phone to photograph bar codes or photograph contracts. There are some roving bank people going to villages, doing banking stuff, photographing it, then bringing it back to the town, connecting their account to the bank and getting it transferred. This is stuff that's really useful for the poor over there, although it started off as a fancy toy in the West.

Getting back to the idea of driverless cars. The weird problem with autonomous cars is, of course, that if they are twice as safe as normal cars and cut the number of traffic deaths in half, still nobody is going to write a lot of cheerful letters to Google and say 'Thank you for saving lives.'

*Yes and no. I spoke about this with Carlo D'Asaro Biondo, who is the head of Google for Europe, Middle East and Africa, and he said insurance rates will probably go down for those who use autonomous cars.*

Yes. That actually leads to another interesting link to my other project, which is reinsurance. I know the insurance people are really scared, because car insurance makes a lot of money. As long as people have accidents independently of each other, the law of large numbers applies and the insurance works and everything is fine. There might be a big traffic accident, but that's just a blip on the radar. However, now you've got your autonomous car. If it's going to crash, occasionally it's going to be due to bad luck, but most likely it's going to be due to some error in the steering system or some software bug. That's common for all the autonomous cars. Google's car insurance is going to be very, very hefty, because, very likely, you're not going to be insuring your car. It's going to be Google who insures all its cars. This is very scary for everybody involved, so I think they're going to try to do something else.

*Yes. I'm sure that people in Mountain View are looking at this.*

From more of an ethical standpoint, of course, if the cars are safer, that's great. We should be using them. If today traffic accidents were a disease, we would be sponsoring many initiatives to cure it. But now, we kind of accept it to a degree, which is bizarre. That's also a scary thing that often comes up in our work here. People accept unthinkable risks because they are used to them.

*One of my final questions I always ask people is this: where do you see us going within five to ten years? It seems that they pay you to think about this.*

Normally, I'm getting paid to think fifty to a hundred years ahead.

*We may not make it that far.*

People here in the office think that we have about a 12 per cent chance of extinction this century. In most disasters, more people are saved medically by the neighbours, not by doctors. Mostly because there are more neighbours than doctors, of course, but people do look out for each other. People actually behave rather nicer to each other in a disaster.

*That is true.*

Going back to the ethical considerations of AI, there are a lot of different reasons that come together to actually keep us working without committing crimes. Some of those reasons are purely for money, some of them are emotional, some of them are our innate morality, but they pile together and produce something that's more resilient. That's also where I think we are headed in making safer artificial intelligence. It is not going to be like, 'Every AI needs to have this little friendliness code in it.'

*A something what?*

A friendliness code, a little routine that makes it nice. That's not going to work.

*People are thinking about that, but they're not sure how to do it.*

Yes, and it turns out to be a very deep problem as Eliezer Yudkowsky has been stating. He is one of the originators of this whole interesting area.

*He's phenomenal. I used his paper on the possibility of releasing a superintelligent AI that is bound and not connected, and he says that it is more probable that, in a wager, the AI wins more often than we do.*[5]

I agree. I wrote a paper where I looked at technical ways of boxing in AIs and it's very, very hard. Based on his paper, I wrote the paper with McCulloch Stewart where we analysed it a bit more technically and it's very tough. That's part of what we do here. We try to show that you can do the stringent academic analysis of what looks like totally crazy questions.

*Totally crazy questions.*

The cool part here is to figure out what that friendliness code would look like. Eliezar's original idea was to program AI to be nice. But what is niceness? What should we be aiming for? Is it enough? When can we say that we have done enough? A lot of current research is going into weird areas of logic and decision theory, trying to find what we can say, and how we can impose some rules on the AI that makes it behave in certain ways, even if it's smarter than us and it changes itself.

This was Stuart Armstrong's example. Imagine you talk to Ganvi, he's a really nice guy and you tell him, 'Here is a pill that will make you a murdering sociopath. Would you like to take it?' It seems very unlikely that Ganvi would want to take the pill, because he likes being peaceful and nice and he thinks definitely the world is not better with him as a sociopath. Perhaps you have another pill that will make someone more altruistic. In that case, Ganvi might say, 'Yeah, that is actually a good idea. I might want to become more altruistic, I do on occasion feel that I am too selfish.' Hence, there are some characteristics you might want to modify yourself. But suppose there is a kind of blind spot that you don't think about at all. Then you're not going to want to update that blind spot.

In Stuart's example you can imagine that there is a guy standing right next to the power button of the AI, and if you set things up just right, the AI might be superintelligent and reason about everything in the world in terms of future consequences, yet ignores the particular small fact that there is a guy standing next to the power button that can turn the AI off. You can arrange things when you originally write the software, so it never cares about this. It would be trapped and unaware of this fact. It is a somewhat peculiar situation, and it's very doubtful that you could put that easily into code, but Stuart has been building more and more other weird things along these lines.

*That's clever. It is clever.*

You can do all sorts of cool things, like you can have various kinds of tripwires, you can actually notice if the AI is doing something outside what you are expecting. At the very least you can figure out, 'I'm wrong about what's going on here,' and hopefully that will tell you something useful about what you should be doing. Then, there are more elaborate things. You might have a being that reflects on itself and what it can say about its future state. This is way beyond me. They're filling the whiteboard with weird logic equations, and I have no idea what's going on. Lobes's Theorem pops up and I think: 'Yes, I understood that for almost two minutes and then it kind of fell out of my brain because that was too weird.'

Getting machines to behave themselves because you put the right coding from the start would be the best. That assumes that all the programs are nice and conscientious, but that is not always going to be the case. The problem might be that you get somebody who just doesn't care or sometimes people say, 'You want other kinds of safeguards too.' I'm interested in more social ones, because one reason we are not all running around committing crimes is that we benefit so much from being integrated in a society. It's cooperative on many different levels. We should also make the machines part of a social system (although it might be more like the internet, which is kind of a social system for software rather than humans). You can construct *soft* ways of controlling things, too. You might actually accept that occasionally the AI goes bad, but that it's no problem. The AI police notice that it's faulty and get it corrected, or at least stop it from being too sociopathic.

*You used a scary term right there: 'The AI police'.*

Who decides what the AI police should be doing? These are issues we need to reflect on. If we design this in the right way we can actually, very gently, guide people and machines to do the right thing. If we do this in the right way, I think we can make better use of it. We can also try to become more deliberate about it. And we should also think about the ethics because this is, of course, all about manipulation.

One of the problems we have with these evolving technologies is that they might empower governments a lot. There is this debate between how technology empowers individuals versus how it empowers various powers-that-be. When you want to do surveillance or facial recognition, the nature of the activity works well with a centralized data structure: a centralized, big computer. On the other hand, we have the network media. When you have blogging, Twitter and all that, people can invent new media and often do. China is struggling to control all the new media people are inventing. It's not just that Google and Twitter are problematic – people are inventing new media, and they don't know how to react to them because the main thing about these media is that you don't know how people use them.

*Going back to the question about the next five to ten years. Where do you see us then?*

In the next five to ten years, you're going to see more powerful technology. We are probably going to invent yet another medium in that period of time. We have an infrastructure that can invent entirely new media relatively easily. Whether or not this is a good thing is hard to tell. Twitter doesn't make sense when you just describe it but it seems to work for some people and for some purposes. You can change the world using new media, using new ways of thinking, organizing or handling your power. Essentially, you're going to see a world that is simultaneously more transparent yet with some very opaque areas that will try to be even more opaque. We are also going to struggle, because the amount of leaks is going to skyrocket. You will have this problem: occasionally, all your really darkest secrets are going to leak out. Somebody who I think was from the CIA said, 'Yes, in the future, all our secrets are going to be on Torrent or Pirate Bay. But that's no problem, because our enemy's secrets are going to be the next Torrent download. You're going to get the Russian and the Chinese secrets there too.' He was okay with that, yet the diplomats were not okay.

*It depends.*

It depends. They are going to say, 'Well, if you have no dirty secrets, you don't need to worry about transparency.' But that's not exactly true, because we want privacy, we want to control things.

*You have a right to your privacy even if you're not doing anything wrong.*

It's very interesting to see how quickly the governments are all agreeing that surveillance is totally right because if you have nothing to hide, then why object? At the same time, they immediately try to safeguard their own secrets when the citizens want to ask for them. This is going to continue. Meanwhile, the automation is also going to make the governments more powerful, because if I figure out how to do something, I can make a piece of software. I can give it away or I can sell it, and now everybody can do that little thing. That's going to transform a lot of activities. Previously, it took a super expert to hack a computer, but now you can download a script and do it yourself. You might still not be very good at covering your tracks, so the Postal Police might catch you. At least part of that got outsourced, and of course software gets better at learning things. It's going to be interesting to look at what skills get automated over the next five to ten years. Some of it is going to be pretty obvious, and some will surprise us completely.

Autonomous cars are a good example. People didn't believe that at all just a few years ago. Now people are thinking it's almost inevitable.

*Anders, thank you so much. It's been a delightful conversation.*

12.

# *Philosophy*
## *(Johan Seibers)*

Although I am obviously interested in how new technology is shaping our times, my normal occupation is teaching philosophy. Fortunately, I was able to meet up with a Dutch philosopher who came and visited the Pontifical Lateran University last year. Professor Johan Seibers is Associate Professor of Philosophy and Religion at Middlesex University in the UK. Before going to Middlesex, he was Reader in Philosophy and Critical Theory in the Department of English Language, Linguistics, Literature and Cultures at the University of Central Lancashire, where he continues to hold an honorary readership, and he is a member of the School of Advanced Study, University of London, Senate House. He also worked for Shell Oil Company, creating future scenarios that could be studied in order to calculate risks the company needed to take.

Even though I showed up an hour late for our scheduled meeting in central London, Johan was kind enough to sit down with me at the Senate House and talk about the philosophical aspects of the new digital age. His reflections are personal and profound, and although at times the content of our conversation can be challenging for those unfamiliar with philosophy, a persistent reader will be well rewarded.

* * *

*What are your thoughts about artificial intelligence? I am going to talk with people at the DeepMind project tomorrow, yet they are very secretive.*

I don't know what their goal is at DeepMind but I think the ultimate goal of AI is to create an artificial intelligence that would be indistinguishable from a human being. There are still huge obstacles to overcome as far as that is concerned. For example, natural language understanding: this has eluded us up until now and I cannot see that they would have access to any kind of insight that we don't have already about natural language understanding, which is not enough to reproduce it artificially. The only real difference today is that we can throw more computing power at it than before. This means undoubtedly we can achieve a greater similarity between human language use and the language processing of machines. I don't quite know and I'm speaking a little bit out of turn but, of course, in linguistics there has been a statistical turn by using enormous computing power and big data to develop a more statistical understanding of language structure; more, that is, than a rule-based understanding of syntax. This is a very recent development in linguistics.

*That is very profound.*

And I imagine that they could apply something like that and attain an impressive artificial intelligence, or at least something that is artificial and looks a lot like intelligence and conscious use of language, which is more than structured symbolic behaviour; it means to be a speaker of a language.

*Okay. We can come back to this theme in a minute. May I ask if you agree with the philosophical distinction between syntax and semantics [in other words, is the structure of a language (such as grammar) different from the meaning of a language, which represents a classical theme in linguistics]?*

Well, I'm not a professional linguist although I did study these things. In terms of my background, I was taught by the famous Dutch linguist, Pieter Seuren, who was one of the founders of the generative semantics movement. He always stressed the organic unity of semantics and syntax.

*Okay.*

The original generative semantics movement came out of the first phase of transformational generative grammar, where the idea was that the syntactic surface structure was a transformation of a deep-level semantic representation to a syntactic representation. Thus, the function of syntax was nothing other than to take a semantic dependency tree and translate it into a linear structure that you can actually voice. This approach was called generative semantics and a few people around Chomsky initially started it. From what I understand, it was quite aggressively cut down by the Chomskians, who soon dismissed the idea of a strong rule-based relation between syntax and semantics. They were supportive of a minimalist programme, as you know.

*I think that Chomsky has since repented.*

I haven't actually followed recent developments. It could well be, but there was certainly a phase where he tried to make the syntactical component completely independent from semantics, from any semantic content.

*I recently read a phrase from Kurt Gödel, referring to his incompleteness theory, in which he states that the irreducible relation between syntax and semantics is a demonstration of the uniqueness of human intelligence. I think he means the presence of the soul, personally; he doesn't say that, but that is my interpretation.*

I would agree with that. I think all of this is very important. For me, what is important in understanding language and in understanding thought, is to understand that there is something that cannot be reduced to a rule-based operation, nor to a statistical pattern.

*Okay. Let's talk about that. I'm very curious to hear what you think about meaning or meanings.*

Meanings represent a very difficult area to think about, but I would like to say that in meaning there is always something that is not generated by us that comes from without. There is something given and meaning is infinite, it is endless. And the fact that these two things are there implies that meaning can never be reduced to an operation that we carry out, and that seems to me to be important. For me, artificial intelligence is, in a way, uninteresting, because it never touches on the question of meaning. It never gets there and, furthermore, it seems to say we don't need it. First of all, I don't see how you can do without it; and I also don't see why you would want to do without it. It would be terrible if meaning were to disappear.

*If what were to disappear?*

This infinity of meaning, its being beyond itself. Heraclitus says no matter how deeply you look into the soul, you will not find the boundary.[1] I think that begins to express what meaning is. Heraclitus is ahead of Aristotle here. His statement does not only imply that 'the human mind is in a certain way all things', it also implies that the structural principle, the measure, of the soul, if I can call it that, is boundlessness.

*I think from a historical point of view you are absolutely right in terms of the apparent incompatibility between a very strong artificial intelligence research project and meaning. Therefore, the initial interest was to deny the importance of meaning. I think as we have become more sophisticated and we have begun to attempt to simulate natural language in artificial intelligence through some sort of interface, meanings are now very, very important. What you said before, that this is extremely complicated, is exactly right; so I'm not exactly sure where I am on this either. However, I do think that the strongest AI projects today, like DeepMind from Google or Watson from IBM, are trying to simulate meaning. I agree with you in that the machine will never possess meaning, but I think it will eventually simulate what we consider having meaning to be.*

Exactly: simulation is the proper term. So, the simulation is based on producing something that, to somebody, looks like something else. I think we can place this idea in very simple terms; for example, in the book *Meaning of Meaning*, by Richards and Ogden from the 1920s[2] . . . Ogden was the translator of the *Tractatus* by Wittgenstein and I. A. Richards is the great rhetoric scholar who wrote a wonderful little book called *The Philosophy of Rhetoric* which contains his lectures from the 1920s and 1930s, in which he says (I've always thought that he was right about this) no matter what anybody tells you, don't believe that anybody can explain how it is that we can have one thing standing in for another. So, the basic symbolic operation, that one thing stands for another, is a function of consciousness, if it is not simply what consciousness is. How is it that there can be an identity between something and something else, which remains another thing? Here we can also see what it means to be a speaker of a language as opposed to a language processing unit. The speaker of the language occupies this ambiguous symbolic space: to speak a word that is not the thing, but somehow the thing and the word belong together and can stand in for each other. For a language processing unit, there are no real words, you might say. There is no *verbum*, no act of speaking.

*Are you talking about a token? Is this what logicians refer to as a token?*

I think it is a completely general problem and I would call it the problem of symbolization. A symbol is something that stands for something else. Without it, we cannot have consciousness. We do it all the time yet we cannot explain how it is that we do this. We can only say we do it all the time; we can understand that it's always happening and it underpins everything. It may well be that you cannot make a machine that does that, but it is an open question and one whose answer will depend more on trying to make one anyway, rather than on speculation. We don't know what reality has yet in store.

*But the machines manipulate symbols, don't they?*

Yes, they are seen to be manipulating symbols to someone who already understands the symbol.

*Okay. You're right. Now, backtracking just a little bit, when you use the word 'consciousness', what are you referring to?*

Yeah [laughs].

*I didn't want it to be an embarrassing question. Personally, I do not know how I would answer it.*

Consciousness is knowing that you know, experiencing that you experience.

*That is a pretty classic understanding of what is consciousness.*

Consciousness is any first-person, subjective state, on this I am with John Searle. I don't see how anybody has ever given a credible suggestion as to how these states could be explained only on the basis of third-person states. It is perhaps even the other way round (something Kant would have said, I think).

*I don't think it can be, even by definition. Searle gives the example of feeling pain as a first-person experience: you know what feeling pain is, and you can also describe that objectively, but those are two different things.*

Yes, exactly. Although AI is concerned with intelligence, not consciousness, let alone self-consciousness, I think AI is based on the presupposition that one day it will be possible that, on the basis of third-person facts or processes, we will be able to create something that is a first-person state. I do not see how that could be done. I think philosophically it raises interesting questions because, at that point, either you become a Cartesian dualist and you say, 'There are two fundamental substances' or you say, 'No, somehow this is possible' and you become a panpsychist, in the sense that everything has consciousness. Actually, that is the line that I would defend, to a very far extent. Consciousness is already there. But intelligence and consciousness, as I said, are not the same thing. Intelligent behaviour, even learning behaviour and the management of ambiguity, fuzziness and ambivalence, may well exist without self-consciousness. There are many examples of it already, and it is an open question what might yet turn out to be possible.

*I would not have thought you would defend that position, but explain it to me further.*

Because it seems to me this would be the only logical option; that somehow this consciousness must be everywhere, latent, or in the nucleus. Consciousness must be there, even in the most basic excitations of the real. There must be already a kind of referral back to itself, in some form or another, in the very act of being itself. I would say that gets us out of a lot of problems that otherwise seem to be unsolvable. Perhaps, many ontologists might say, the price you have to pay for that is far too high, because now suddenly even electrons are at least potentially conscious. But it is just another way of stating the identity of being and thinking, a good old metaphysical principle. Being is just that which is thought, or known; thinking is thinking of being. The two are the same and yet different. We can understand this in terms of a divine consciousness or in terms of subjectivity and I think we can also make it amenable to a broadly materialist ontology.

*You're absolutely right: that's very consistent.*

Some philosophers have come very far trying to think in this way; for example, Alfred N. Whitehead in his great contribution, *Process and Reality*, in as much as he takes experience as an ontological process, or reality as a self-actualizing event, an occasion or act of experience, in which, in a way, this consciousness is actually already right there in the constitution of the event: it is in the basic form of the real.

*That is incredibly interesting. I don't know if I would want to go there. Let me throw this at you: could consciousness be an attribute of a being that has a certain complexity? So, in this way, we would reserve consciousness to some, but not to all.*

Yes, and also Whitehead would certainly say that only what he calls *hybrid prehensions* can have the kind of consciousness that we would consider consciousness and that is probably true. It's almost a statement of fact. This does not mean that the principle of consciousness (which in my view is not something complex but simple, namely, something stands for something else) cannot be present in a much more primitive way.

*Can I ask you where you think consciousness comes from? In a way, you've already answered this using panpsychism, latent in all of reality.*

Yes, then the question becomes: 'Where did all of reality come from?' Why is there something rather than nothing? A question often misunderstood today by limiting its scope to the presence of a material universe. It is a confrontation with nothingness, perhaps as an impossibility, or at least a 'rather not'. Leibniz formulated this question in this way for the first time. In his words the 'rather' is already there. There is a will-intensity to existence, it is not mere presence. Understanding why it may not be possible for us to give a definitive answer to these questions is more important than translating their meaning or offering an opinion about them. I don't think it is helpful to say, 'I believe this or I stand here'; it is important to understand what the reasoning behind these positions is. I have, personally, come to a view where I accept that there is a very strong argument to be made in philosophy for the existence of necessary Being; that must be seen as the cause of the rest. I would like to say there is obviously contingent being; contingent being cannot be fully understood without a reference to necessary Being because if all being were contingent, there could be a situation in which there were nothing, but that is a contradiction, or rather when we try to think of that we discover that our concept of 'nothing' is a limit-notion, perhaps even a pseudo-concept. And so affirming a necessary dimension in being seems reasonable: it can't be such that there is nothing at all.

*That sounds like Thomas Aquinas.*

Yes, this is a paraphrase of the Third Way. I may change my view on this again, but at this point in time I think that from many viewpoints that is a very sound way to think about this problem. What do you think?

*Obviously, I teach at the Pontifical Lateran University and it is nearly impossible to be a Catholic philosopher and not agree with Thomas Aquinas. If you're speaking specifically about consciousness, I'm not sure where I am at the present moment. I may have an idea which might sound contrary to Catholic doctrine and let me just test the waters with this idea. It is very controversial and so if you prefer not to comment that's fine. I'm taking a large risk by mentioning this but here goes. It refers to artificial intelligence. If consciousness is the result of a sufficiently complex computational process, then AI will achieve it. This is a big 'if' and I think there are people working in AI who believe that as long as we get the power and the complexity going, then it will achieve consciousness. I disagree with that point of view, but I do think there is a reason that people hold it. The controversial position I want to take now is the following: if we managed to create a sufficiently complex machine, God would endow it with a soul and it would then become conscious. Hence my provocation to you.*

That is a very interesting idea. Let's take a few steps back and thank you for placing this question in front of us. Karl Rahner had a similar idea, and Church officials often did not like him; at least in the beginning and also looking back, such ideas were too controversial before Vatican Council II and even after it. Rahner, I believe, had the idea that there might be extra-terrestrial intelligence and there might be Christ events on other planets – that God would also have sent his son to whomever these intelligent species are. I always thought that that was a wonderfully open-minded idea, and that it does not take away anything from the unicity of the Christ event, but that singularity is not in conflict with the idea that something happens more than once.

> *That is correct. That is quite a tenable proposition, although we would have to consult a theologian to be sure.*

Something similar might be the case with the creation of beings with souls. There might be, even if we stay strictly within a theological or even Catholic perspective, a latency in nature, the machine with a soul, which is providenced in God's plan but whose factual existence depends on us making a machine that can house a soul. It is an inverted Frankenstein story in a way, or a 'homme machine' in a sense in which De la Mettrie had not conceived of it.

> *Well, I don't know what that being will be.*

It could be something that we don't fully understand yet.

> *Which is precisely my fear. The philosophical or scientific rationale behind a position like that (and I know that these are unexplored waters) stems from considerations concerning in vitro fertilization: we create matter that is offered to God for 'ensoulment' and this usually occurs within the womb of the mother.*

Yes, but it does not have to be only there.

> *Now with the technology we have, it does not have to occur in the womb but can also happen in a petri dish. This is what Aristotle called the predisposition of matter (which usually occurs between a*

*father and a mother). We now can do this independently from them and we still achieve a human being. No one doubts that a baby born this way is human, though I think the question may have arisen at the beginning; but now it's obvious that they're completely human. Let me take it one step further: what happens when we create artificial sperm (with which they are now experimenting)? So, now we have something we have created and by using a donor egg, we can fertilize that and we have a human being. So what happens when we create an artificial egg and artificial sperm?*

Ultimately, you could build it up from as deep as you can go. It would involve a more complete science, as you take a bit of matter and build it up. You would not even say that you are playing God, because you are simply manipulating matter in the way that has been given to us to find out how to do it. But it is perhaps even more interesting to ask why we are driven to these types of fantasy, what does it say about the human beings we know exist that they are so interested in the question of mastery over nature, so much so that they create their own destruction in the process, if we are not careful. I think it is fundamental to the human relationship, to being as a whole, that we can experience it as a gift, as something that we have not made but that reaches out or relates to us, and vice versa. Maritain spoke of a basic ontological generosity that defines what existence is. Even in our fantasy of mastery it comes up, in the idea that God will endow matter with a soul. What is that other than saying that in the light of the absolute, something is made part of a totality without being reduced to it. In a metaphorical sense, everything we make is endowed with a soul. Now, if you are right and God does his part at that point in time, assuming that it be possible, you could then ask the question: does he want to do it? Does he have to do it?

*Well, that's a very good question. In the Middle Ages, this was actually a question that was presented concerning a child conceived from a sinful act (like in the case of rape or a couple out of wedlock) because such a child was conceived violating, in a sense, God's law. Well, no one holds that position any more because those babies were just as human as everyone else; they were just as human as babies conceived in love (which is the way God wants it), so God himself is bound to the laws of nature because he has chosen to be.*

He has chosen out of freedom and in a way can no longer interfere with it and thus follows the laws of nature. However, I also believe the laws of nature can change and we can see that they are slowly changing or we just don't know.

*We would have to ask Him.*

There is no rational reason for us to say in advance that the laws of nature are unchangeable.

*I'm not sure I follow you on that last notion.*

It is really an inductive inference that the laws of nature are the way they are, as opposed to different; as, for example, the gravitational constant could be different and because it could be different, there is no reason not to suppose that it might have been different at one point.

*I agree, but I don't know what the point of that is.*

The point is that you said God is bound to the natural laws of nature, so I would say maybe not. A 'law of nature' might be just another statistical generalization. I think we should not forget to hear the oxymoron in 'natural law'; it is almost an ironical concept, a *physis* which is *thesis* at the same time. Nature does not need the law, and the law is not natural. The concept, and the easy use we make of it, hides a metaphysical embarrassment and, perhaps, more of a peek into contingency and groundlessness than we often care to admit.

*Okay, I see how those two arguments can be made.*

But what we are speaking about now is not necessarily the same point. You can hold that position and it does not interfere with your view on the possibility that God would endow a machine with a soul.

*Here is my dilemma: if this does happen, we might witness an atheist AI developer who would claim that he created consciousness. So the question you asked thirty minutes ago is the most important: where did this consciousness come from? I think, as Catholics, we need to start addressing this. Instead of saying the machine will never be conscious, we need to examine all possibilities.*

I think that is a very important and a very true argument. It also links to a point made by Pope John Paul II concerning evolution when he said that Catholic thinkers should not try to find problems with evolution theory, that we should not repeat the errors of the Galileo case.[3] We should learn from the errors of the past. The Pope insisted that the soul is created each and every time, and that is something which cannot be accounted for in evolutionary terms but we need to make it work while embracing evolution as a biological principle wholeheartedly. This is the official position of the Church and I think there is a parallel between that way of thinking about evolution and your idea about sufficiently developed artificial intelligence, such that consciousness may arise.

*From a hypothetical point of view, how far away do you think we are from something like this?*

As I said before, I don't see it happening. It would be so miraculous to my understanding if an artificial intelligence engineer would be able to make a machine like that; yet, if it happens, I would be happy to take your view and say God must have put something in there!

*I see on the one hand you are open to the idea but on the other hand yours is a strong criticism of AI.*

At a philosophical level, I don't understand how further complexity can lead to the light going on, but I am open. I do not think there are conclusive *a priori* reasons that you can adduce and say 'It is not possible'.

*I think Nick Bostrom*[4] *refers to this in his book, where he talks about a missing link. I think he is on the same page as you are. Just because you augment complexity to the* nth *degree, this does not give you consciousness and he agrees. But if we find the missing link, it may happen.*

I am open to that and yet I cannot see what it would look like. It is not a link, the metaphor is misleading because it does not incorporate the main point here: that complexity and consciousness are on different levels. It's not a broken chain. Perhaps bridge is better. We need bridge builders for this task and maybe there is more human creativity that needs to go into it than we think. Maybe we have to sensitize ourselves more to the life in all things before we can see it also in machines. AI is interesting also because of this: we are putting ourselves on the line, in more ways than one.

*Precisely: we don't know.*

In my view, it would have to accomplish a kind of *creatio ex nihilo* in a way, and in that sense, I can understand why you say what you say.

*Okay, can we talk about the future?*

Yes, let's talk about the future.

*Can you tell me more about the future, and what you did at Shell? I understand that you were a member of their original team that projected 'future scenarios' in order to help plan effective risk management.*

I've always been interested from a philosophical point of view in the future. I tried to understand temporality and how it relates to our own existence. To me the future, the fact that there is newness, seems to be central to what it means to exist, and yet the discussion of the future arrived relatively late. People don't usually think about *being* and *thought* in terms of the new, in terms of the future. If you look at most views of knowledge, they have some sort of relationship to the past. In Plato, knowledge is remembrance and that goes all the way through to Hegel: knowledge is of what was already there; or *essence* in Aristotle pertains to that which something 'always already was'. The idea that the openness of the future might be a basic, real dimension of being, that being in other words is to some extent radically undetermined, is not very old.

*That is quite metaphysical.*

It is the idea that there is a structure; there is a reality and thinking is only finding your way back to it, which in a very strange way contrasts with an experience of life that leads into an unknown future, in which something new might come about or in which we create things that were not there before; the realm of creativity and surprise, of the radically new.

*In previous conversations, you mentioned that you have worked with other institutes; and you've worked on future planning. My question is that it would seem obvious that we need to reflect on this. Let's just take one example: investment. You invest in something and you expect to have returns in three years. You obviously need to calculate what the future holds and it seems so obvious that we would be much more involved in this kind of reflection.*

We sometimes make a distinction about what 'investment' actually means. Where does the word come from? Today, we use the word investment only to indicate a potential return and outcome and output rather than its more original meaning where we 'invest' something with a particular quality or a particular purpose or meaning. We have an instrumentalization of investment, a term originally referring to the act of clothing someone in the robes of an office. Now when we hear the word we first think of financial returns and then we think money and only then do we realize that actually we invest a specific object with meaning and so this word carries within it human agency, the possibility to create, change and make new and the human capacity to take things into one's own hands. This human activity of investment is something we need to think about in new ways to understand what it really is, how it is about having a nurturing, agentive relation to a potential future rather than simply one of calculating what my return will be and managing the uncertainty of the future experienced as fate.

These are two fundamentally different attitudes: the first one has to do with the future as an open horizon of possibility. How can I relate to the fact that the future is open? The other one is a colonizing attitude towards the future: how can I close down the future now sufficiently already so that I know what will happen in three years' time and can use this knowledge to my benefit? That attitude of closing down the future is really a denial of futurity in a way, rather than the first one, which is an opening to what the future has in store and taking up an attitude of responsibility for the future rather than its exploitation.

*I guess it depends on what your priority is.*

Of course, there is a place for both of these things.

*You won't make a lot of money unless you in some way direct the course of the future.*

Yes, but even in the context of business planning it has become very important to make people aware; this is one of the first things that we do in workshops for scenario planning. The first difference we talk about is that between scenario thinking and predicting the future.

*I'm not sure I understand that notion because I see those two as very close.*

Yes, but they are actually different things because predicting the future means reducing uncertainty: based on what is happening now, we can extrapolate that such and such is going to happen in the future (like predicting the weather), whereas scenario planning is thinking in these terms: based on the drivers that are present today, what are the alternative futures that might be and what are the knowns and unknowns?

*I was going to ask in terms of your experience: aren't corporations doing that kind of thinking?*

They are doing that kind of thinking and they are doing it to get a better handle on their environment; you might say the temporal environment in which they exist.

*I was thinking about Facebook, for example. I'm sure they have teams working on these kinds of things.*

And of course we did the same in Shell. Actually, Shell invented long-term scenario thinking along with the RAND Corporation in the 1950s and 1960s. They invented scenario thinking and the goal was always to come up with alternative ways of looking at what the future might look like and to use that whole set of alternative scenarios to calibrate or to judge business ventures. They asked, 'Will this particular business investment or this business decision that we want to make hold up in each of the scenarios?' The key phrase was scenario robustness: will your business decision be a good one regardless of the scenario that materializes with respect to the business environment?

*I guess you could always be wrong too; did you come across that?*

Yes, of course. I was at Shell Headquarters when 9/11 happened and we had been working on scenarios for two or three years and no one had come up with that specific scenario.

*That's amazing, but who could have foreseen such an event? How would you explain the rise of the Islamic State, for example?*

That is a difficult question.

*This is like a pseudo-nation, it's a caliph.*

It is not a nation state by any means. It uses the demise of the nation state to describe itself in what is in fact a geopolitical region.

*They receive taxes from their people; they have state education and a state hospital system.*

If it goes on, there will come a point at which it will be recognized as a state, and like many other states the bloodthirstiness of its founders will become a matter of history, diplomatic relations will be set up, and so on. I can see that happening.

*What can you tell me about the future of humanity with regard to digital technology?*

I've been thinking about this theme and I've recently been doing a lot of work on the future of education and collaborating with people who are working on the future of education. There we see the impact of technology in a dramatic way. We can see that the whole idea of a curriculum is changing. The idea of a classroom is changing and what the teacher does is truly changing under the influence of technology in ways that are atypical. When you are presented with new media, people understand the new medium in terms of the old medium that they already know. For example, when we started to write emails they looked like letters. It took several years for people to realize that emails were not letters. Marshall McLuhan has written a lot about this; for example, with his analogy of the rear-view mirror: it's as if we were driving into the future looking into the rear-view mirror – we understand what is coming in terms of the past, with which we are familiar after all.

*But is there any other way we can do it?*

It appears there are two modes or mentalities of knowing: tying something back to what you already know, or letting go of what you already know to understand what is new. McLuhan held that artists do the latter. They live right on the shooting line and that is why people say artists are ahead of their time, because they are the only ones who live in the present. And so today educators can learn from artists. They have a great opportunity to give new meaning to Freire's remark that education is at once 'conscientization', politics and art. As education is emancipated from the dominance of certain media, new forms of education can arise with new media. But we can also focus again on those aspects of education that have to do with encountering others, and so in a way with an emancipation of media themselves. We can become aware in new ways of education as an act. But predicting technological development is a very tricky business. We just don't know what lies five years ahead. We really have no idea.

*I think today we can identify several of what you referred to as 'drivers' and then we can, to a certain extent, predict where those may take us.*

Yes, we try, but there is also an agency involved, isn't there? There is something we can do (like education): we may be tempted to succumb to an ever greater instrumentalization of the educational process, in which education becomes more and more just the transferal of skills or the training of skills, assessed by measurements and tests. Or we can search for a more diversified approach in order to get away from the idea that one size fits all. Look at the school as a production belt of diplomas or the industrial model of education: you take the raw material in; you put it through a standardized process; you do a quality check at the end and get the specification out. Even though that model is dying out, it has not disappeared. It's dying out because our technology allows us to be much more diversified in how we manage that process; but education still happens within the parameters of an instrumental view of what education is. There is also the opportunity to use technology for the development of new ways of educating that actually move away from that old idea to explore and experiment with new ways in which education can be personal formation, holistic and humanizing.

*I'm so glad you mentioned that because at my university we have just hosted a convention on the concept of 'bildung' and formation like 'Paideia' in ancient Greece. The ironic thing for me was that the only avenue we did not discuss was the digitalization of formation. We talked about Renaissance man, we talked about the ancient Greeks . . .*

These historical examples can inspire and we can learn from them. But if we get stuck singing the reactionary blues, we've lost it, especially in education, which is so centrally about youth, change, blossoming and opening up, a lifelong youth. We can and should salute Renaissance man, but we cannot now go back to him.

*I guess I didn't have the strength to say that to my colleagues.*

Why not? Because we tend to think about technology or the digital revolution as something that is an enemy of humanization. This is where I would say future studies can help: to find ways to think about digitalization not as a threat to humanity but as an opportunity to explore avenues that we may not have even known about. This is what an open mind to the future can actually show you.

*So would that be your personal position?*

That would be my personal position and if I may add one thing to that, I think this has ramifications far beyond education or artificial intelligence; I think it touches on our understanding of religion as well. I think it is necessary today to take up the idea that Whitehead began to explore in *Religion in the Making*, which was first published in 1926. His idea was that change and development are much more intrinsic to religion than we tend to think. I think in most institutionalized religions there is a kind of harking back to a past, a fetishization of origins and tradition, whereas that runs completely counter to the message of Christ that God is a God of the living and not of the dead. Christianity has a forward-looking direction embedded into its very essence. It seems to me that clinging to the past, to a fixed identity, is often related to fear and security, but embracing the new can only be done with courage and hope. Religion as a human reality has both elements in it, because they are both part of what it means to be human. But I think the genuine life-giving newness in being that all religions (not just Christianity) have always sensed and known about is especially important today. Religion has something to offer there that is unique. This openness is much more important than the closed, all-encompassing meaning that religion sometimes wants to pour out over everything, causing unspeakable suffering in the act. And this openness belongs to religion, by right and from the start. Sometimes the word religion is explained by reference to the Latin root *religare*, to tie back to the divine, or also to past tradition. We could think of the human orientation on the future and on the full, unknown gamut of its potentialities perhaps as 'proligion'. Our ties forward to the future are much deeper than our usual notion of 'progress' implies, and there is a place for the heart as much as for the head in them.

*I think that is because we tend to be control freaks, we want to be in control.*

The attitude of wanting to be in control is at the root of the technological attitude.

*In what way?*

Heidegger wrote a lot about technology as a means of control. I think he missed an important point, which one of my professors pointed out to me once. He says the thing we often overlook when we think about technology, also Heidegger, is that if you don't get it right, it doesn't work. If you are fixing a car, and you don't do it right then the car doesn't work; it doesn't drive. So, doing it right is not up to us, it's not up to us to specify how it is to work because there are properties to matter. The idea that with technology we are in complete control over a completely plastic reality is not true. We have to be attentive to what reality shows us of itself, especially in our technological relation to it.

*One of the issues that people are bringing up now is nanotechnology.*

Nanotechnology is a good example of what Ernst Bloch called 'alliance': there is a possible alliance between nature, the human sphere and technology, or at least we can think of the relation between technology, humanity and nature as a relation of alliance rather than as a relation of exploitation and control.

*Actually, that is a good point on several levels, not just in macroscopic terms.*

Yes, it also applies to the real nuts and bolts of what technology is. There is more to be said about technology than 'it is just control'. I do take the idea that comes back to the Frankfurt School really, that Western culture has been marked by a controlling attitude towards nature and also towards our own natures. The orientation on the future has got something to say about this. It wants to free us from this old, oppressive aspect of our own psyche and our own behaviour. I think it is deeply ingrained in almost all cultural institutions. The root of it is fear. Today, now that fear is becoming ever less obvious in some ways and is seen as something to be managed and got rid of, the control to which we commit nature, others and ourselves has become extreme. We have not liberated ourselves from this basic feeling, instead we repress it more and more, and as a result our attempts at control become ever more intense and unaware. It is as if we don't dare to practise alliance instead of control out of fear that there will not be enough for everyone.

*I think it is so deeply ingrained that I am more cynical than you are. I can see that it is much more plausible that the course we are going to take will be indicated by the 'drivers', or specific interests, especially the interests of the powerful. Rather than open-mindedness and an almost deference towards the future and what the future may hold.*

Deference or gratitude – an attitude that pulls us into the direction of the religious sphere. And not just that. It also brings us in the vicinity of what justice means. Justice does not close one off from the future, but keeps open the possibility that the future might be different, it keeps the future itself open. That is a core aspect of justice. It is what the heart does when it forgives, for example; there is no forgiveness without a future or without hope.

*I think the future will be much more humane if that is the course that we take.*

But I see your point: there is little indication that this is the course we will take. Look at Nick Bostrom and the world of the transhumanists: if anything, they are creating a new discourse of privilege; they are creating a paradigm of the haves and have-nots. How much can you afford to enhance yourself?

*Do you think that this has already begun?*

In a way it's as old as the world itself and so nothing new, but the scale of the dimensions to this age-old drive to improve and distinguish ourselves, that is something we really need to get our hands around. But what do you propose then?

*I'm trying to speak with intelligent people like you and see if we can't perhaps defy the status quo and the historical trends that have marked humanity. We did not speak about war but we could discuss that. Pope Francis has said that World War III is already in act but only in a piecemeal way. Look at all the violence in the world. Look at what happened several months ago in Tunisia.[5] The city that was attacked in Tunisia, Sousse, was built by the Italians. It's on the Mediterranean Sea, a beautiful town, with five-star hotels and discotheques. It is a very Western city and that is why they found so many British and other Western vacationers (which is why they chose Sousse as a site to attack). As the Minister of the Interior stated, 'No country is at zero risk'; he basically said this could happen anywhere. I think he's absolutely right.*

I think the Pope was very right when he said we are already in a state of global war. This is something we've been speaking about for many years: the spread of war and the theme of warfare tactics used by actors who are not nation states, like a pandemic, a state of total war, not as permanent military violence, but a smouldering everyday that erupts unexpectedly, now here, now there, a state of permanent terror which has no outside any more.

*I am somewhat cynical but also hopeful because I am a priest; I am hopeful that we can harness the powers that would lead us to a better future and a more humane one.*

Obviously, the Church has a role to play in this and I hope that the Church sees itself as a shepherd of people, into an open, free future, let us say 'greener pastures'. But I think the role of all churches or religion is diminishing, even though there has been somewhat of a return of religion in quite a remarkable way. If you look at the 1970s or the 1980s, a lot of people in Western Europe (maybe more than in America) would have thought that religion as an influence in public life was about to disappear completely; yet that has not happened.

*But is religion synonymous with spirituality?*

No, it is not. There is faith, there is spirituality, and there is religion.

*Unfortunately, what I see is a trend towards conventionality of religion, especially in the West. The people who are still religious, I think, are so to a large extent because they are conventional. There is a value to that, even though I personally am not that way. I understand that many people are religiously conventional and they value that. I think there is an increase (especially among the young) of living a spirituality that has no part in organized religion.*

And that is very strong. I see it among my students. I see it everywhere. But it is an open question whether so much spirituality can sustain itself, or whether, as Scholem teaches us so beautifully with the example of Jewish mysticism, the spiritual life needs a framework within which it can flourish, which it also always transgresses and disrupts, but without which it is almost too vulnerable. Religion has to open its doors widely for the most exotic forms of spirituality. It risks excluding the most beautiful flowering of our religious existence if it is any less than welcoming and willing to be a student of spiritual teachers. The rise of spirituality in Western culture is, at least in part, a call to change addressed at the churches. Here Rahner, whom we've already mentioned, proved well aware of the future himself. He held that mysticism would be the core of future religiousness.

*In terms of digital technology I am also consoled by a tendency I see among my students (tell me if this is your experience too): as the technology becomes much more pervasive (they've grown up with smartphones, with Facebook), they tend to use it less.*

Yes, it becomes less of a thing to do. Yes, that's right. I have begun to notice that too. There have been years of being so overexcited with the new, yet now that is wearing off a little bit. I think that applies also to the internet at large. For a time, it was such an explosive change in our information and the way to view information that it took us a while to find our bearings. But I think we are doing that more and more, and that gives hope. The fact that technology wants to disappear into the background is hopeful.

*I think people are more interested now in the quality of their communication and the transferal of information than simply doing it, than simply communicating. I have seen that for example in YouTube in terms of going viral and receiving millions of hits in a short amount of time: those videos are really well done, and they are really well thought out too, there is something which excites the imagination. People are not just clicking on anything, whereas before perhaps they did. People are much more demanding.*

Yes, we are finding out that we have to manoeuvre in intelligent ways when publishing things on the internet, but also that our understanding changes of what it means to be an author, a publisher or a member of an audience or public. It's interesting to think about, and a good example of a scenarios question: what will internet use look like in ten years' time? I suspect that a lot of it will recede into the background of our lives, at least I hope that it does. I think it is sad that in public spaces everyone is obsessed with their little machines, and eventually people will have had enough of that.

*I think sooner rather than later. I think ten years is too long, although I do not know. I was waiting for the train at Gatwick and on the platform everyone was looking at their smartphone. And there is something psychological to that: the need to feel connected. Yet I am optimistic. Let the machine do what it does well and let me live as a human being. So, if my smartphone can take care of a lot of things for me, like mundane issues or things that must be processed, the exchange of information . . . I don't need to be involved in that, as long as it is reliable.*

These are the lines along which we should think about the future, about technology and its implications for the future because in this way it becomes a creative task, it becomes a question of how we actually shape the future that is ours. To shape to a large extent rather than sitting speechless in awe of an unknown and unknowable reality, or gazing at signs in the sky trying to figure out what is going to happen. The hopeful attitude becomes something more than the question 'Do the indicators favour pessimism or optimism?' Hope becomes moral, an active commitment to the proposition that change for the better is possible. Bloch says we do not have the right to be pessimists. We certainly do not need philosophy if we are pessimists, for in that case the situation takes care of itself. Bloch's remark may be a rhetorical formulation but it points out something important when thinking about optimism and hope: they are not things that you can base only on the evidence.

Hope is always in the face of the hopeless, otherwise there is no hope. And that is why hope is such an important category when we think about the future. You cannot understand what futurity is if you don't talk about things like hope and despair. They are not secondary effects that come once you realize there is a future. They are the ways in which the future presents itself. That is why these things are so important to discuss.

Kant uses a wonderful analogy in *Dreams of a Spirit-Seer*:

> I find no attachment nor any other inclination to have crept in before examination, so as to deprive my mind of a readiness to be guided by any kind of reason pro or con, except one. The scale of reason after all is not quite impartial, and one of its arms, bearing the inscription *Hope of the Future*, has a constructive advantage, causing even those light reasons which fall into its scale to outweigh the speculations of greater weight on the other side. This is the only inaccuracy which I cannot easily remove, and which, in fact, I never want to remove.

This 'inaccuracy' is ingrained in us in a way that appears as a mistake from a certain perspective. Do you have sufficient grounds to have hope for a better world? Then you have to say 'no'. But there is a kind of mistake in us, a brokenness, and that brokenness, that lack or absence, is what provides the hope of the future. For it, even lightweight reasons will do. Hope for the future is the inaccuracy that makes us whole. We cannot accept the way the world is: we have to try to make it better. This attitude towards humanity, history and futurity does not mean that the angel of history sees no catastrophe. But it can help us to understand a bit better the depth of our investment into the future, which goes as deep as human existence itself. I think this is the way we need to think about the future. We have to do away with the technological and colonizing attitudes towards the future that seek simple solutions to big questions. (But this does not mean that we should do away with technology, far from it: there is also a future for technology.) For these solutions in the end mistake means for ends, and make ends – our desire for immortality, for love, for happiness – into means, means for more colonizing technology. They serve to repress our awareness of the meaning those questions have: questions of the alliance of human, nature and technology, questions of being at home in the world with others, questions, ultimately, of what Bloch called 'the strongest anti-utopia', death. When we think about the future, we have to think about what hope is, what anticipation is, and how these may open up a future to us.

13.

# *Nuclear Instruments*

## (*Troy Anderson*)

It was a mutual friend who introduced me to Troy Anderson. Troy's background is engineering and he now works at a large nuclear company called Canberra. Canberra Industries is a subsidiary of Areva, which builds nuclear instrumentation and radiation measurement systems, from hardware to software solutions. Areva is one of the largest multinational corporations for nuclear power plants. It is present in many countries and employs over 45,000 people worldwide.

Even though traditional nuclear power plants are not very 'digital' (because the fundamental technology arose before the digital age), Troy's insights on the impact of digital technology are very helpful from a scientific point of view, as well as giving us an insight into an industry that continues to be a disruptive influence in the energy sector, many decades after it first entered into common usage.

* * *

My name is Troy Anderson. I've been working for Canberra for almost twenty-five years now. Prior to that I worked as a contractor for the navy. I went to college for physics, mathematics, and then electrical engineering. I went back for electrical engineering. Since I graduated from college, I've been basically using all of those skills in the various jobs that I have had. Canberra has probably been the most successful place where I've been able to take everything that I've learnt and employ it, which is really nice.

*What is your specific role in Canberra?*

I am Senior Chief Architect. Basically, I'll go out and look at the new technologies, bring them in-house, whether they're software or hardware, and come up with different ideas and employ them and see how successful they are. If they work very well, I'll try and sell them to the organization through presentations, showing them the pros and cons, any of the cost-benefits. If the company buys into it, then I'll typically go off and start providing training throughout the organization. Canberra's scattered throughout Europe and the Americas.

> *In a conversation I had with Massimo Morichi, former Vice-President of Research and Development for Areva, several items concerning the Fukushima disaster were analysed. He mentioned some ideas which you might want to comment on as well. One was that the power plant survived the earthquake, which was very, very strong and frightful. But it didn't handle the ensuing tsunami very well: the diesel motors that were going to be used for back-up electricity failed.*
>
> *And the other thing that he mentioned was the theme of cyber security. Obviously, the security of a nuclear plant is extremely delicate and important. Nuclear power plants have very, very tight controls in terms of the software. They are not allowed to be connected to the internet. They have very, very strict protocols that imply also physical restrictions, not just software restrictions, in order to enhance the security. These are obviously very sensitive subjects – maybe you can say something about that.*

With respect to how power plants use modern computer systems, they typically will have two networks. One is the business network, which is secure, and then the other one will be the instrumentation network, which is very separate from the business network. The reason why is because those systems are very, very old and security was not a huge concern back in the early 1980s through to the 1990s, so those systems typically aren't as secure as what you have on your computer systems today. They have a separate network that's disconnected.

The other thing that power plants have deals with safety zones: for anything that's in a safety zone (an area that is very critical to monitoring a person's health and safety, for example), there won't be any computer systems at all. It will all be analogue-based, it will not be digital-based. The newer systems that Areva has been trying to put into the newer power plants (and they've been kind of forging ahead with a lot of these ideas) are introducing digital systems. At the same time, typically we see that industries today don't want to change because they have tried-and-true systems which have been around for twenty or thirty years. They don't want to suddenly introduce these new technologies that really are more susceptible to cyber security threats.

I don't know how they're arranged in Europe or what the definitions are but in the United States they call them 'still bubble'. There are different levels of safety that software has to go through and the solutions that Canberra has been providing (and I believe even Areva too) have been at the highest safety level, and that's why they're not down at the areas which would be more exposed. These solutions are completely safe in that regard.

*Well, that's consoling to know. Just yesterday I spoke with a founder of a cyber security company in Miami, Florida, and they do business all over the world: he said there's no such things as 100 per cent secure software.*

That's true, that is true. Everything today can be hacked, it's just a matter of finding someone that's smart enough to figure a way into the systems.

*He used to work for the National Security Agency, so he had a lot of experience, but you can't mess around with nuclear power plants. Even 99 per cent secure is not secure enough I think: you're right it has to be 100 per cent.*

Yes, and you're going to have the human factor involved. Human beings are fallible when they're writing the software for these systems; the originator is the human being, so mistakes are going to be made. What's interesting today is everybody's expectations for hardware and software solutions; in fact hardware solutions are really software solutions. Everything is completely controlled by the software. Those expectations are increasingly getting higher and because of that the software is getting more and more complex. With the increased complexity, you're basically introducing less security.

*Exactly, that seems less consoling.*

Yes. It's funny because I worked in the software industry for many, many years and when they introduced online banking, I was adamantly against it.

*Really?*

Everybody, all my peers and friends, they were like, 'Why?' And I said, 'I work in this industry and I can see the flaws on a daily basis, and I see this as being very insecure.' This was back ten or fifteen years ago and now, all of a sudden, there have been these reports on how many banks are very compromisable today. I've been saying this to friends and family: 'See? This is what I was telling you about, it's just been hidden, people didn't know about it.' Now, smart people are trying to find interesting ways of gaining access to banks and stealing money and making a profit.

*Do you consult your account online?*

I do not [laughs]; no way.

*Old-fashioned. Talking about software, I don't know if you're familiar with the Stuxnet virus: it's a virus that the US government created along with the Israeli Secret Service, I believe. They managed to get it into an Iranian nuclear power plant and it made the centrifuges go crazy, destroying a lot of them, while the whole time the administrator of the network was led to believe that everything was fine: the computer was saying it was all in order. Are you familiar with that event?*

Yes, and I didn't think it was a power plant: it was a nuclear processing facility. I believe the intent of that virus was to basically slow down (if not stop) the processing of nuclear material that could be used for either what Iranians are saying – a power plant – or what the world believes – the creation of a weapon of mass destruction. Who knows what the truth is. So, yes, I am familiar with it and I also know that because it was put out on to the internet: it's being used in very negative ways against us and probably other Western countries.

*Yes, do you want to give me your evaluation of that? It's like shooting ourselves in the foot, isn't it?*

It is definitely shooting ourselves in the foot. Although I understand the reasons behind it, I don't necessarily share those reasons, or agree with them. Any time you create software that has some sort of malicious intent, you're asking for trouble. With software forensics today anybody can reverse engineer it, and whether it's going to be used positively or negatively, that person could use it for whatever purpose they want to. You really do have to be careful with what you're pushing out on the net today. That also goes with the information you're putting out, like on these social media sites. Or when you're using your phone, you should monitor the services that you actually have enabled, and those that were enabled without you even knowing about it.

One of the things that I've started to do (because I really do get concerned about Big Brother these days) is when I get my phone, I disable all location services. I only enable them when I need them, along with a variety of other services. I only enable services when I know I'm going to use them for that period of time and then I completely shut them down. Bluetooth is another concern: I don't leave it wide open, I completely shut it down.

*Wow, you're scaring me a little bit here, but it's making perfect sense.*

Well, I didn't mean to scare you [laughs].

*The ability they have to monitor us is astonishing, and it's a little disconcerting, I think.*

Yes, it is.

*So, Massimo Morichi mentioned that you're like a future trend-alert looker. It seems like the company pays you to think about what's coming next. What are some of your thoughts in terms of this digital revolution that we're living in? Eric Schmidt calls it 'the new digital era', where the first one was Microsoft and the introduction of the PC. Now we're seeing many, many more millions of people coming online. I think that artificial intelligence is producing some pretty wild stuff also.*

It's interesting that you mention this, because twenty years ago when I first entered the computer world I started working on work stations, old Solaris boxes, HP Unix, and those were the cutting-edge systems of the day. Then, in the 1990s, I witnessed the PC revolution and I embraced that and all the Microsoft technologies as well. When we got into the 2000s, I started to see the train moving towards smaller and smaller processors with more power and it became more of a mobile revolution, or an embedded revolution. That's where I think we're peaking these days: we have really pretty much exhausted that area. I get the sense that social media seems to be on the increase. I would speculate that it's around 80 per cent height level, and so it's going to peak out soon. It was very interesting to see how that had an influence around the world. Especially with Facebook and the Arab Spring. All the communication was done through social media sites. I was very surprised by that; it just completely blindsided me, I didn't expect that kind of influence. I remember when Facebook was coming online: it just didn't seem like much to me [laughs], but it's surprising to see just how much of an impact it's had.

*I think you're right though, I wonder if we haven't peaked already. A lot of my students are not using Facebook any more – these are students in their twenties. It seems like more parents are using it, so the kids don't. I'm seeing that people realize they don't have enough time, because if you have maybe 1,000 friends and you go on Facebook, all you can do is follow everybody and what they write. You can be there constantly for twelve hours and not do anything else.*

I think it's like a fad. People really jumped on to it quickly and it took off, and now people are saying, 'You know what? I don't really have a lot of use for this.' I think that's where solutions like Twitter and others have taken off as well, because they're just little blurbs of information that you can put out there, instead of having to have these exhaustive discussions on Facebook.

You mentioned artificial intelligence, and I do get concerned by some of the things that we say we're going to do in that arena, because we're really trying to take the human factor out of a lot of the computer control, and have the computers or the software make the decision.

*Yes, that's true.*

That is a very concerning area because you really don't know what's going to happen there. It may be just the fear of the unknown, but given all the movies that we've seen, and the books that we've read about the negative side of AI, it is an area that you need to cautiously step into. We as humans tend not to want to be cautious about things, we tend to just kind of dive in and say, 'Well, we'll deal with the repercussions later.'

*It's interesting that you should say, 'We're not sure what's going to happen.' I agree with you, but I think that a counter-argument could be made. Computers are run by rules, their systems are set up by rules, and human beings don't like to follow rules sometimes, whereas the machine will. This may even be a plus for us to have the decisions being made by a rule-based system.*

The problem with that, however, is the following. Look, for example, at the military. The military is really very ruled-based as well. When a soldier is trained, he is told, 'Hey, you obey the officer no matter what he says,' but at some point you have to allow for the human factor which says: 'Is this the ethical thing to do?' Machines (at least right now) do not have the ability to make that decision. They cannot tell the difference between, 'Okay, is this rule ethically right, or is the rule something that I don't need to worry about and I can just blindly apply it?'

*The military is actually trying to implement ethical rules and embed them in their unmanned vehicle systems, also known as drones. Again, you can make the argument that those are just other rules. For example, the drone is about to fire and it encounters an innocent baby next to its target, then the rules should kick in, 'Do not fire', because there's an innocent baby that's going to die. I guess I would say you're absolutely right: a machine cannot have an ethical sense, but it can have ethical rules.*

I guess you can say that. Back to the drone thing. Today's drones are completely human controlled, they're not autonomous machines.

*I hope you're right.*

They may want to get to the point where they're completely autonomous, but that's where we're going to run into some of these issues with the complexity of the situation. Will the artificial intelligence be able to make the right decision based on how complex the situation is? For example, you can add in these additional ethical rules, but I think it's going to be more of a trial-and-error thing. This means that as we're developing these systems, we're not going to think of all the possible cases right up front. Mistakes are going to be made and then we're going to have to kind of 'grow' that intelligence from there.

*Where do you see your own field going within the next five to ten years? We're going to still have nuclear power plants, no?*

Within Canberra's domain, it's a very slow-moving market. What I see (and what I've pretty much been leading the company through) is a transitioning from things that typically run on a desktop computer and putting more of the intelligence within the device. In the process of doing that, given there's more connectivity options today, you will have the devices talk amongst themselves, and hence you can make even smarter decisions.

When it's processing the data that's coming in, it can go and grab information from nearby sensors and say, 'You know what? I think it's this but these other guys are telling me it's really this, so let me augment that information before I present it up to the end user,' and either make the decision for them, or basically let them make that decision. That's where I see Canberra really heading. I've been pushing us in that direction for the last ten years and we're on the verge of getting very close to doing that within the next two to five years, I would say.

*Can you give us some concrete examples of this kind of technology?*

I can mention a few. Today we're actually getting into the area of environmental monitoring. We're producing an instrument that will sit out in the field and acquire data continuously, twenty-four hours a day, seven days a week. As it's doing that, it's analysing it and making decisions like, 'Well, gee, I see an isotope here, so do I need to generate an alarm?' That's just one device. If you put up a whole bunch of them within a vicinity where they can communicate with each other, you could use it to get directionality information. For example, 'I see this thing moving so it could be a plume cloud overhead because the power plant has had an accidental release of some material.'

*What technology would the sensors use to communicate with each other?*

Today, we've been going down the path of Bluetooth LE, and Wi-Fi, and so basically you're either maintaining a connection, or just making a transient connection to each device. We've looked at various RF solutions as well for mesh networking. But we haven't really stepped into it because there are some limitations. Those limitations that we've encountered are related to securing the network, because you definitely don't want anybody tampering with the data.

In terms of Bluetooth LE and Wi-Fi, you can pretty much guarantee no one's tampering with the data, as long as you're taking advantage of those facilities.

*Would it be wrong of me to extrapolate out a little bit and say, we could start putting sensors all over the place? You will know what traffic is ten miles away through your smartphone; you will know what the temperature is in the next city over; you will know what the air pollution is, etc.*

I see your point. Actually, I was discussing this with an associate of mine several months ago, about maybe embedding something as simple as a really small detector within every phone that you have today, so as people walk around the city they could actually be acquiring information that could be set up through a central server to narrow down where there may be a terror threat.

*Okay, not sure I follow you. I followed you on the whole cell phone thing, but I didn't understand the jump to the terror threat.*

What we were discussing is developing an ASIC, basically a chip with all the smarts of what we do as a business, within it, and then going to some of the popular cell phone manufacturers (like Apple's iPhone, or some of the Android ones, like Samsung) and trying to have them put it within their phones themselves, as they're manufacturing them.[1] As people walk around during the day with these phones, we can be acquiring data in the background without you knowing it, and sending that information off to either a central government site or something else. Some of these solutions we've already seen today, but they've been more like attachments for phones.

*Isn't that a little creepy?*

It is very creepy, but think about the GPS that is in your cell phone today. From discussions with a few friends of mine that are in the cell phone industry, they said that they were forced to put GPSs within the phones by some regulation, or regulatory editing (not sure if it was the FCC), because they wanted to be able to locate the phone if you had an emergency, say. You just dial 911 on your cell phone and they would be able to locate you. That was something mandated by the government, which I was surprised to learn – I didn't know that.

*I didn't either, but I do know that they have locating devices in the phone, yes.*

We were thinking: 'Well, why not try to embed this within the phone?' These were just like futuristic thoughts which probably will never happen because we're in such a narrow market that we really wouldn't have much of an influence, unless there was some sort of government regulatory agency that was pushing it on to the cell phone industry.

*Actually, I can see it happening.*

I wanted to at least mention that during Fukushima, when that happened, there were some really smart people that came up with very small Geiger counters. You could hook these devices up right on the front of your phone, through the ear jack. They had a little lab that would acquire the data and display it to the user. People over in Japan, if they were concerned about their radiation levels, they could buy this for around $300 and attach it to their phone and turn their phone into a little Geiger counter.

*Yes, that sounds cool, I would do it.*

It was definitely an interesting product. In fact, my daughter was taking a college course, and the teacher that was teaching the class said that one of her students created the app and he is now a millionaire.

*What are your thoughts on technology in the classroom? One of my students, who is a pre-med, told me that she had to study the different protocols for various drugs. It's pretty complex apparently, although I'm not familiar with it myself. She said it took a lot of effort and a lot of time. Then she started an internship, and she was on the ward and the doctor said, 'Oh, we have to administer this drug, what are the protocols?' She knew them because she had studied them. But one of her colleagues, a girl next to her, said, 'Oh, there's an app for that,' and she whips out her iPhone, she punches a button, and there are all the protocols for that specific drug. What's the use of studying if there's an app for it?*

You're not always going to have access to that information. Just because you have a cell phone today doesn't mean it's going to always be there and be reliable. It can range from any number of problems, plus when you're in a critical situation, again, you might not have access to it. It's always best to use your brain instead of a machine, at any point. Plus, there are a lot of skills that you can lose; for instance, handwriting. Everything we do today is on a computer system. When I sit down to write something it's a chore to do it now, whereas it used to be really easy. When I was at school I'd write papers and solve problems, all just on paper. It's very, very different today and you might lose some key skills that you really need to carry through life.

It kind of goes back to when we were younger and calculators were coming out. I'm forty-eight years old so back in the day I had to do everything on paper, and then when calculators started becoming more cost-effective (meaning you could buy them fairly cheaply), people were saying the same thing – 'Oh, well, you're going to lose all your skills' – and in the end we actually didn't: we still maintained those skills. I'm not quite sure where the line is drawn in respect of the modern technologies.

*Yes, I'm not sure either. I agree with you, but I think that, more or less, we're always going to have a reliable smartphone with us. You're right, maybe your battery has run out, maybe you don't have any access, maybe you forgot your cell phone, but I think these are exceptions, aren't they? I think that we're generally pretty reliant on our cell phones, aren't we?*

I would say you're definitely right there. I actually find during the course of the day, I'm always using Google to look things up, or using an online dictionary to check spellings, just to make sure that I spelled things correctly. In the past, what would I use? I would either have to use my memory, or I'd have to go to a library to find that information. It is very convenient today to have everything at your fingertips.

*Do you see that trend becoming even more and more profound? Even more at your fingertips, faster, more reliable, more universal, capable of doing more things?*

I see there probably will be a day you'll have a chip either connected directly into your brain, or somehow wired to your brain, where you can gain access to information very quickly; instead of having to type it into a computer you will be able to see it right in your own mind. I know they're trying to go down that path; in fact there have been several articles online, and even videos on TV, discussing that. It's definitely interesting, they are interesting ideas. It can be scary in a lot of ways too because of some of the issues we see today with software, and even again, as we go back to cyber security, what could happen.

*Let's imagine that we had a chip, like the prototype which Sergey Brin (the co-founder of Google) allegedly already has at Google X. If this chip were proven safe, you could take it out when you didn't like it, it didn't have any collateral effects and it basically gave you Wikipedia in your head, hardwired to your central nervous system, would you get one implanted, Troy?*

I actually wouldn't, no. It's interesting now that we're going down this path in our discussion; I didn't mean to go down this path but it's bringing up some concerns that I went through about a year ago when I ended up having an operation to have a defibrillator put in . . .

*Oh, I didn't know that.*

I'm a very physical person, I've been exercising since I was a kid. I'm still exercising now that I have the defibrillator as well. The doctors happened to catch a condition, luckily, when I was being monitored. My heart rate is extremely low so they wanted to monitor me for twenty-four hours. They caught this condition and they said, 'You're going to have to have a defibrillator put in.' So I agreed, and I said, 'Okay, let's just get this over with.'

After the operation I went to have them run a diagnostic test on the defibrillator, and they didn't tell me what they were doing. They just rolled this computer system in and they waved this magic wand in front of my chest, and they walked across the room and started playing on the computer. I'm sitting there reading a book and the next thing I know my heart rate is increasing and I'm feeling very uncomfortable, and I look at the lady and I say, 'Are you doing something?' And she's like, 'Oh, sorry, [laughs] I'm increasing your heart rate by three-fold now.' And I'm like, 'Well, I feel it.' She goes, 'Are you uncomfortable?' and I said, 'Yes.'

*[laughter] By the way.*

I wasn't really uncomfortable because of the heart rate, I was uncomfortable because I had no control over myself.

*That is amazing.*

This made me think about the ramifications: 'Well, gee, if I pop this chip in, yeah, it will give me access to all these things, but what else could happen?'

*Absolutely. That's a great story. Dick Cheney has a pacemaker for his heart, and in 2013 he had his doctor turn off the Wi-Fi function it comes with, because he was afraid of getting hacked.*[2] *He was afraid of somebody manipulating his pacemaker. It's a way of possibly getting assassinated.*

Yes. The same thing is true with the defibrillator. They have this little wand, which unlocks the device, and then they can communicate over some secure protocol. I don't think it's as simple as Wi-Fi, but they have other protocols that are out there. That's very, very concerning. In fact, after I had the operation I started reading about some of these things. I read the Dick Cheney article also.

*You saw that too, okay.*

Yes, and there were a couple of other articles that I read related to the potential of unsecured protocols. Even though they say they're secure, they can be compromised. Granted, I'm just an everyday Joe, so no one is going to just walk up to me and try and kill me for no reason. But I can understand, in the case of Dick Cheney, there could be a reason for someone to try and do that.

*Yeah and who knows, maybe somebody realizes that you're valuable for an industry, that you hold secrets that they want, and they could kill you. I would hope they don't succeed, but they may actually want to do you in.*

When I was reading about all of this stuff, there were several papers that were released within the last year that showed some students at various universities who had found out how to compromise these devices. They said, 'Well, this is the equipment we use, even though ordinary people would not have this happen.' The point is that they are compromisable. Especially for me, it's quite unsettling because I have this device inside of me.

*Yes, it is. One of things that I often ask people is the following: do you think that your job could be done by a machine?*

Actually, I don't think so.

*That's good.*

I think once we get to the point where a program can write a program, I might be able to say, 'Well, maybe my job could be done by a machine a little bit.' The things that I do, like being able to research various technologies and see how you can put them together to create new things – it's going to be a while before artificial intelligence has the ability to be very creative. In the solutions that I've seen today, it doesn't sound like there's this creative gene within that software yet.

*I think you're right. But when you say 'a while', what would be a ballpark figure for this type of breakthrough to occur?*

Maybe twenty to thirty years.

*Exactly. If you think about it, that's not a long time in the big picture. Your daughter will be middle-aged right, at that point?*

Yes, I'm sure she'll be getting close to that. I wouldn't say that you're going to have artificial intelligence permeate everything. I remember when I was growing up, they were saying that computers were going to take over everybody's job (that was back in the 1980s) yet we haven't seen that happen. We're seeing some areas where some mechanical drones are doing some manufacturing for us, but it's not in large quantities.

*Right, you mean like in Detroit?*

Yes, Detroit, and then you also have Amazon with their drones for fulfilling orders.

*Yeah, but you're right: humanity hasn't been decimated by this. It's like agriculture. The world used to be based on agriculture. Now only 2 per cent of the world's population is working in agriculture, because the machines do it all.*

Yes.

*You seem very realistic. Previously, you mentioned movies like* Terminator *or* Blade Runner. *Am I right in assuming that you favour a point of view that reconciles humans and machines, in the sense that we're going to use the technology to advance our own interests as human beings, and not be destroyed by it?*

Yes, I definitely would say that.

*What advice can you give us, from your perspective?*

I definitely would say to use the technology cautiously and try to be aware of potential vulnerabilities in the technology. Just as I was mentioning the use of cell phones today: I understand the various ways you can attack the devices, so I just try to use the services when I actually need them and then disable them. Try to be smart when you use the technologies and protect yourself.

*Do you think that people today are doing that or not?*

People are definitely not doing that. I think that they really don't realize the vulnerabilities in the technology they use. It's something that the average person just hasn't been informed about, or they don't really know how bad the threat could be if it happened to them. It's exactly like this whole thing with the defibrillator. My wife was sitting in the room and she was chuckling when I was saying to the nurse I was uncomfortable. After the nurse left I said, 'No, I'm serious. This is not funny. I had no control over myself. Think about that.'

*Oh yeah, it's devastating.*

When I put it that way, my wife said, 'You know, I never really thought of it like that,' and I said, 'That's the point.' I even told her about the possible threats with her cell phone: 'Just be aware of what you're allowing, or putting out to the public.'

*Exactly. Am I right in that this device is actually saving your life though, isn't it?*

It's basically a safeguard. I have a condition which is called a tachycardia event: it's where your heart can start pumping from the bottom up, and the defibrillator is to prevent that condition from happening. It happens to me when I overexercise. Sometimes I go out and do some long cycles and it's very taxing on my system. I'll do 100 or so miles and when I get back at the end of the day I won't properly rehydrate myself and that's usually when the condition occurs. So far, since I had the surgery, I've been very cautious about things. I make sure I drink a lot of water all the time and I make sure I don't really push myself as hard as I used to. That's one of the things the doctors emphasized: 'You're getting older, you're not the same kid [chuckles] you used to be.'

*But there are trade-offs, you're absolutely right. I think that you're probably glad that you got this device, but at the same time it's a little unsettling when it takes over.*

Yes, yes.

*Troy, this has really been an interesting conversation. Thank you for talking with me.*

Thank you, it's definitely been interesting for me as well.

## 14.
# *Mechanical Engineering*
## *(Bill Shores)*

Having heard that one of the engineers on the original team for Motorola that invented cell phone technology was willing to talk with me, I wasted no time in contacting him. His name is Bill Shores: a soft-spoken, middle-aged engineer who lives in Arizona. Although he owns a cell phone, he hardly ever uses it, and often discourages others from using one. For someone with such a prominent background in the industry, this attitude struck me as unusual, and I was curious to find out more.

Bill's career coincides almost exactly with the invention, development and mass distribution of personal cell phones. It has been said that if a Martian were to visit our planet, he might think that the devices were part of our human anatomy. And there could be no better person to give us an insight into this particular aspect of the digital era than Bill.

* * *

*So, who is Bill Shores?*

I guess I am just an engineer, probably unlike some of the other impressive people you've interviewed before. I went to Michigan State University for undergraduate work, got an electrical engineering degree, got a Master's at Arizona State University. I've been working primarily in the telecommunications field for nearly thirty years. Once I got my Master's, I went to work for AT & T Bell Laboratories. I was working in deploying intelligent networks worldwide . . . it was a lot of fun. Actually, the first customer I worked with was in

Italy. I was working with Italtal and SIP, to deploy intelligent network services in Italy.

*Can I interrupt you? Just briefly, what do you mean by 'intelligent networks'?*

Intelligent networks, back at that time, meant something very different to what it does today. That was back when landline services were very prevalent and mobile services were much less prevalent. That included free phone service, virtual private networks, all using landline phones.

*And how many years ago was that?*

It was actually in the late 1980s when I got my Master's when I started doing that.

*Wow.*

So, I was working in that area, I was actually working in telecommunications standards. I was travelling and working at ETSI quite a bit, which is the European Telecommunications Standards Institute. I was also working at ITU, the International Telecommunications Union, working on standards. At that time, those services were very new and innovative and important. With the boom of mobile technology, however, those services went by the wayside because the billing models associated with mobile technologies were basically that you just pay one price and whomever you call doesn't really matter: you're paying for airtime.

*But at the time, that was cutting-edge technology.*

Oh yes, it was pretty important at that time. And then there was a lot of convergence for a while with mobility services, the boom of cellular networks and intelligent networks and the convergence of teleservices. A lot of that went into the MAP [Mobile Application

Part] and CAMEL [Customized Applications for Mobile network Enhanced Logic] standards – I'm not sure if you are familiar with those or not.

*No, can you briefly explain that?*

MAP was basically the networking protocol that was used for deploying GSM [Global System for Mobile Communications, the standard adopted by almost the entire world as the way to communicate via cellular calls]. It provided most of the mobility aspects; it's what allowed cell phones across different network providers to communicate.

*Wow.*

And a lot of that technology is based on similar technologies that were used in intelligent networks that I mentioned earlier. You're familiar with roaming services: you can take your cell phone and roam to any country you want.

*Yes, but it costs an arm and a leg.*

You can blame the providers for that.

*I do.*

So I had worked in cellular-type services when they were just starting, starting from the convergence of intelligent networks. Shortly after that I received a call from Motorola because they were looking for people with my type of background to work on the Iridium system – I don't know if you have ever heard of the Iridium system?

*No. Doesn't ring a bell.*

It is a low earth orbit satellite-based system that was conceived back in the early 1990s and actually was a spectacular success from a technical standpoint, but a spectacular failure from a business standpoint. The idea was that they launched a network of satellites that rotated around the earth and they are still up there working; believe it or not, I just recently started working back in that programme because there are very few people who have background in that area. In any case, the idea was that they wanted to provide a service based on GSM, the European cellular standard, but it provided the ability to communicate anywhere in the world. So rather than using cell towers, it used roaming satellites as the cell towers.

*Which are a little further away than cell towers.*

Yes, so you have the delay issues, and some other issues with the lead times which were necessary to deploy satellites, launch them and everything. It is a network of sixty-six active satellites that rotate around the earth. The term Iridium came from the element iridium, which has seventy-seven electrons rotating around its nucleus. They actually came up with a way to reduce the number of satellites, reducing the cost and covering the earth with sixty-six. I forget what the name of the element with sixty-six electrons is, but it was a horrible name so, from a marketing standpoint, they just kept the name Iridium.[1]

*That's funny.*

The network is still up and running, but the problem that they ran into was this: they came up with the idea to provide ubiquitous service around the world, so anybody could go anywhere and use their phone and be able to communicate. But they didn't foresee just how quickly the cellular technology would bloom. You may recall that in the 1990s it was unbelievable how fast GSM technology just exploded across Europe and similar technologies in the United States. The problem with that, of course, was the fact that now you had these very expensive satellite-based technologies that were competing with relatively inexpensive terrestrial-based technologies. At some point it became very clear they would not be able to close the business case on that, so from a business standpoint, the Iridium company itself went bankrupt, but the assets were sold to another company and it has been up and running for twenty years. In fact, they are launching new satellites next year and will be replenishing the network.[2] What they are focusing on is more niche markets, more vertical markets, things like shipping, cruise lines, places where there really isn't any competition and there's no ability to communicate (like in the middle of the ocean): airlines, shipping, freight areas. They are doing well now, because they are focused on vertical markets.

*One of my students went on a cruise from Barcelona to Rome and she said they had Wi-Fi the whole time. So, that means there was Wi-Fi on the cruise liner, and I can understand how you could get to a central server on the cruise liner, but how does your signal get back out if not through satellite?*

If they are using the Iridium-type system, there are services that allow for concatenating multiple channels. So they can provide what they call 'Broadband service' but it's not broadband in the sense you'd think of with Wi-Fi or terrestrial. You can provide an internet-type service but it would be at a relatively low speed, something in the order of 64 kbit/s.

*That's really slow.*

Yes, so in terms of using video or audio, it would be horrible. But if you were just doing web searches, I'm sure it would be okay. I've never actually used it on an ocean liner. The new satellites that they are in the process of designing and launching have substantially higher bandwidth. That will be improved. It could very well be that the ship your student was on had some other alternative: a communication line that maybe wasn't even going through a satellite. I don't know what the ship lines use.

*It certainly can't be traditional cell towers: they're out in the middle of the ocean.*

It depends on where the cruise was going and how far out from land they were. If they are not too far offshore, they might be able to communicate directly with some towers in whatever port nearby. Much like a plane can communicate with airports directly. In fact, the Iridium system is used as a backup, only for those cases in which the planes can't communicate directly with ground stations.

*Let's go back to your experience. Can I ask what you do now?*

Right now, I am primarily back working on that Iridium system. As I mentioned, they are launching new satellites, so they are investing several billion dollars to replenish the network and upgrade it. Given the technology that they have, you can appreciate that there's probably very limited expertise in the area. It's sort of a unique situation because even though the basics of the technology are similar to cellular technology or GSM, it gets very complicated in the sense that you have to deal with the idea of the cellular network where not only is the person with the handset moving, but the cell towers themselves are moving. It's very complicated in terms of making sure that as satellites orbit the world, they are continually making contact with ground stations. It adds a new dimension to the problem. I am working on that, primarily based on the fact that I'd worked on the system before and there is limited expertise in that field.

*Your professional career has really coincided with the onslaught of this digital revolution. Am I right?*

You could say that. We used to joke about it. You met my wife, Nancy. She actually has a degree in computer science and she worked in the field for about five years, until we had our twins, and she hasn't worked since then. The funny thing is that it wasn't that long after she retired that the real boom of the World Wide Web came into being. And that was what really mushroomed all these technologies and related issues. It wasn't that long ago when everything changed from the time she had worked. In just a handful of years everything was based on web-type technologies which didn't even exist five years before.

*If I can remember correctly, Nancy mentioned that you have some patents.*

I actually have eight or nine. I'm not sure, I would have to do a search. Patents are interesting things. Some of them are much less impressive than others.

*Okay. And in what field do you have yours?*

Most of mine had to do with a project which I didn't even mention in terms of background. After working on the initial Iridium system, I was working on a variety of other technologies and products in Motorola. Most of my patents had to do with a gateway system that was providing a bridge or an ability to interoperate between SIP-based networks (Voice over IP-type networks) with cellular-type networks that had 'push-to-talk' technology.

*What does 'push-to-talk' mean? Like a walkie-talkie?*

There was a technology in the United States that became very popular. The company was called Nextel. It was basically a cell phone that also had a dispatch-type service. It could actually work with virtual private networks. Try to envision a cell phone that had a button on the side with which to be able to communicate directly with the contact, much like a walkie-talkie.

*I'm following you.*

But it wasn't like you had to be in the same local area as the person you were contacting. You could be in completely different cities or even countries and it would work. That actually evolved into push-to-talk service because it was wildly popular in the United States. I don't think it ever caught on in Europe, although there were standards that were working on it. We were using a Voice over IP technology for multi-media technology. I don't know if you have ever heard of SIP – Session Initiation Protocol?

*No. I have heard of Voice over IP. I think that is what Skype does.*

That is correct.

*And I think that is going to be the downfall of telephone companies eventually. It's already impacting them to a great extent and the trend is to use it even more.*

Voice over IP is interesting. For a long time, my colleagues and I used to joke about it, because it seemed like every year they would say, 'VoIP is going to take over next year.' It was always next year. And this went on for about ten years.

*So, I guess I'm wrong!*

No, what I'm saying is that Voice over IP has been around for a very long time. Every year was going to be next year, when it would take off. Eventually it did. So, even now a lot of the landline phones are using VoIP technology.

*I was just going to say I'm convinced that landlines are becoming less and less common, much rarer. The reason I am thinking this way is because the monthly payment for my landline has continually gone down. This is weird because payments always go up. I think it is because they are using VoIP technology to do international calls.*

Certainly. In the United States, with the younger generation, I don't think people even get landlines any more. I rarely use my landline phone. The only reason we have one is because Nancy's mother has hearing problems and she has trouble picking up a conversation when

Nancy calls her on a cell phone. So, we primarily have the landline phone because it's a clearer signal when Nancy calls her mother.

*That's funny. I know the following questions might sound a little generic, but let's see what area we can explore. Our generation is going to be unique because we will be known as the bridge generation: we grew up without cell phones, without internet, without DVDs. I recall when VCRs came into being. That's how old we are. Our children do not know a world without internet or without cell phones, without Google. Some of my students ask me during class, 'Why should we learn that when we can just Google it?' Do you think we have a useful perspective? Is there something that we can give to a generation that will be completely digital? Just think what Eric Schmidt says in an interview: 'I'm glad I grew up in a society without Google, when you could still make a mistake. Today, you can't: it's all recorded, it's there for ever.' I know some students who are suffering from low self-esteem, some adolescents who have difficulties because they are sending risqué photos through WhatsApp. Once you do that, it's there for ever. We didn't have the same experience.*

It's funny you mention that. I was just talking about something like this to one of my daughters earlier this week. I completely agree with that perspective. There are two primary reasons that I have avoided ever getting on to Facebook or anything like that, and one of them is that there seem to be risks with things being taken out of context. Not that I am worried about risqué comments or anything like that; it's more an issue of things being taken out of context. The other aspect is the whole idea of this arm's-length interpersonal communication. I don't want to get into a mode where I spend so much time online and wasting time.

*Yeah, yeah. Do you see that as a potential disadvantage to the digital generation?*

I do, as a matter of fact. I do see it as a disadvantage and I struggle with it. What I see is this whole connected world, which frankly is disturbing to me, mostly because it seems like kids are so leery about actually speaking. They don't really talk to each other, they text instead of talking. Kids these days (at least in the United States) don't seem to be enamoured with Facebook any more, mostly because their parents are on it. They find their alternatives. But the bottom line is: instead of having real interpersonal relationships, which means talking face to face or even by phone, they just don't. And that seems very disturbing to me. There is something you miss from the interpersonal standpoint, and it just seems like a bad thing to me. Kids use the internet for 'infotainment' and entertainment, and that seems to be quite destructive. People are frittering away time, following celebrity gossip, not necessarily doing terrible or sinful things but certainly wasting time, worrying about people that they don't even know and what's happening in their lives. I recall a friend of mine one time telling me this quote that has always stuck with me: 'What we invest in the famous we rob from ourselves.' When I see people following the mindless Hollywood nonsense and being so absorbed in what's going on in celebrities' lives, I feel like, 'What a waste of your own life.'

*That's true. That's certainly a downside. How much do you think that is caused by the ease of communication, like following people on Twitter, or is it a sign of a kind of immaturity or superficiality on our part, independent of the technology?*

Honestly, what do I know? I'm just an engineer and engineers tend to not have a great perception of social things. I think it's more of an immaturity, although I need to think about it more. All these new interconnected technologies just make it easier. I think people have done that for decades anyway, it's just much easier to waste time today.

*You seem to imply that the jury is still a little out on this. Do you think the technology which we are developing and that is getting into our lives almost in an intrinsic way is, on balance, good or bad?*

I believe that the technology is intrinsically good or at least not bad. It goes back to the comment about immaturity or a lack of interior life. There is a strong temptation to do things which go against what you should be doing. From my own life, I think it's a wonderful thing. You mentioned your students saying, 'Professor, why should I learn this when I can just google it?' I don't necessarily buy into that. Anything I search for, I take with a grain of salt. I don't necessarily trust what I see on the internet. Google is a fantastic thing because I can't remember everything I ever learnt thirty or forty years ago. It's amazing to me how quickly I can relearn things that I used to know, which would take me probably days if there weren't a Google. I can relearn things that I knew a long time ago within an hour. It's stunning to me how quickly I can recall and reapply things. At the same time, I've also run across things on the internet where I know enough from having learnt the basics in engineering to say, 'Wait a second. This is not correct.' So, when you have students who ask, 'Why do I have to learn this because I have Google,' how do they know that what they are reading is actually correct? I do see that attitude in children here and it's wrong. They want to take the quick route, they want to get the answer, they don't really want to understand how the answer was derived so they can figure it out themselves, which I think becomes dangerous.

*That's an excellent insight. I haven't really heard it stated that way. They want the answer without having to figure it out themselves.*

Personally, I tend to be very curious about how things work.

*That's what engineers are supposed to do!*

It took me about twenty to twenty-five years to realize that I just have to accept that fact. For the longest time, I was an engineer and everyone wanted to push me into management, saying, 'You should be doing more.' So I went into management for a while and I really didn't like it. I struggled with it and then at one point I realized:

'Why have I been fighting this? The way my mind works is like an engineer. God made me to become an engineer and that's just how it is.' It worked out for me. I've got that natural curiosity and so I don't necessarily believe what I see: I want to work through it and make sure I really understand it. Kids always want to take a shortcut, and you hope that at some point they realize these shortcuts aren't doing them any good. The downside is that the Googles of the world make taking shortcuts much easier.

*Like anything, it depends on how you use the technology. Let me pick your brain on the following, and I know it is difficult to say, but where do you see us going within the next five to ten years? The reason I ask you this is because some people have even speculated that the future will be 'Us against the machines', they are extrapolating certain trends that are happening in an exponential form and they've come up with some pretty wild scenarios. Where do you think we are going to be five to ten years from now?*

Maybe the best way for me to answer that is to say that I oftentimes have to force myself to not think about it because it actually scares me.

*Oh. Okay.*

Just look at the things people are doing in terms of artificial intelligence. The technologies are there to make great strides, but I don't know if the morality is there to keep it in check. That's what concerns me. I look at all these wonderful ideas about, for example, immortality (and what Ray Kurzweil is working on), working with biotechnology and nanotechnology. I look at that and I say, 'That may be wonderful but, from my own perspective, I don't want to live that long.' This is what disappoints me: I think, 'Well, it seems pretty obvious to me that people who are interested in living for ever probably don't have much of a faith life.' That may be an oversimplification but it is true. The unintended consequences always scare me.

*And you're thinking that this is a possibility within the next ten years?*

I hate to put a time frame on it, but I can certainly see a lot of these things happening in my lifetime. Whether that's ten years or twenty, I guess that's up to God. I think there is a convergence of multiple things: the technologies are getting there to apply super-intelligence; you can see artificial intelligence getting to a certain point and I'm sure you are well aware that they are running experiments now with driverless cars. Google and Apple are both doing that. I look at that and I think: 'That's a terrible idea.'

*Why is that a bad idea?*

Multiple reasons. Besides the obvious potential safety issues, I get the sense that we've got a large swathe of the population that seems to be perfectly happy with taking the lazy path. I don't necessarily apply this to driving. And by the way, when I talk about these things I do so from being immersed in the American culture. I know what I say is not applicable worldwide. I just see so many people who are perfectly happy to let machines and computers do their thinking for them. I am a big believer in the idea that a lot of spirituality and dignity comes with work. People are doing themselves a disservice by allowing themselves to become so lazy that they really don't want to do anything any more.

*That's kind of depressing, actually.*

That goes back to the point I made earlier: I try not to think about it.

*I think that it's unfortunately inevitable. That's where I'm coming from. That's why I'm doing these projects – the first book was sold out within two months and I think that's because people are interested in reflecting on these subjects. Like the conversation that we are having. It's strange that we don't talk to each other more about this stuff. Everybody has a smartphone, everybody navigates. I use examples in my homilies that deal with things like 'spam' and everyone understands exactly what I'm saying. Many of them ask, 'What is he doing speaking about these kinds of realities instead of speaking about Christ?' We just take it for granted that we all have this technology, we're all immersed in it, but we really don't understand what it means.*

Yeah. And I agree with that. I think it is important. I am glad to hear you say that because one of the things that concerns me is that there is far too little time in which people are focused on their spirituality and interior lives. They fritter away time on the internet, chasing celebrities, which just detracts from that. So, I would love it if there were somehow something to bring people back into focus. This technology is great but it needs to be used wisely – I just don't see that happening enough. While I believe that GPS and geolocation services are inherently wonderful things, it really bothers me that companies like Tinder exploit these technologies to promote impromptu sex hook-ups amongst strangers. Not to mention that it drives me crazy when people don't want to be bothered with learning how to read a map – what will they do if/when their technology fails?

*You read my mind. I was going to ask you what advice you would give, what kind of suggestions you would have, because this is happening and it's going to happen.*

My advice would depend on what you are talking about. I will interpret it from the broadest angle possible. One piece of advice, certainly in terms of these wild ideas about immortality and artificial intelligence, would be to be very cautious. People tend to underestimate the impact of unintended consequences.

*History is full of examples of that.*

Unfortunately, I'm not sure people pay attention to history like they should. One of the *nice* things about technology now is how quickly it moves; one of the *risky* things about technology is how quickly it moves.

*Exactly.*

This would be my biggest piece of advice: just be cautious. It's a wonderful thing but there are always unintended consequences. Some of the things we are working on now could be absolutely devastating.

*You're coming at this obviously from personal knowledge; this is your field, you are an engineer, you know how this stuff works. I think you can see where we are headed. So, that's pretty valuable advice, I think.*

I wouldn't give myself that much credit as being in that sort of future-looking work that is cutting edge. A more simplistic perspective gives me concern. There are things I can do now just by virtue of allowing a computer to do calculations and run through many scenarios so quickly, which enables me to make decisions that weren't practical twenty years ago. Although this is wonderful, it shouldn't be left unchecked. With the amount of computer power now, if people can make decisions about things or proceed with something based only upon the ability of a computer to churn out results that don't take into account factors that are important, there is always the possibility that bad decisions will be made.

*Yep.*

As more and more gets put on artificial intelligence, I get worried. Not that I work in artificial intelligence, but I believe there needs to be some non-machine making the ultimate decision.

*You're starting to sound like science fiction, but it's no longer science fiction: you're absolutely right. Let me just end with something provocative. You've heard about the co-pilot of the Germanwings plane that went down in the French Alps. He purposively downed the plane. If we did not allow human intervention, the plane would have landed safely in Düsseldorf.*

I know exactly where you're going and I've thought about that as well. I am so leery of just putting machine overrides in place to avoid human error. This is going down a path that makes me uncomfortable. Machines could begin to override people's decisions. In this particular case, one of the things I thought was that it would be really nice if there were a way for the tower or a human being down below to override the co-pilot's decision. I understand what you're getting at, it's just that sort of world gets me nervous. I've seen the movie, *I, Robot*, the one with Will Smith where the machines make all the decisions. Granted, that's far-fetched, but having machine overrides starts taking you down that path.

*I don't think it's that far-fetched. The example I gave you was obviously a trick one in the sense that if you were on that plane, you would be screaming for a machine override. In the last couple of minutes, the passengers started realizing what was happening. You are right in that if we give the power to the machine then we are going to have to accept the consequences of that.*

Yeah. Enough said. I would prefer an option in which you would have the ability to have a human command the machine to override.

*Right, so you keep a human in the loop. Let's conclude on that note. It's been a scary conversation. I was thinking that this would be a little more uplifting in terms of an engineer with technology. What has occurred in the last hour is that we are reaching the same conclusions as we're going through this. We have to keep praying and reflecting on this.*

If you had asked me fifteen or twenty years ago, you would have gotten different answers; or if you were to ask me now and I was fifteen years younger, I would have responded differently. As I've gotten older, I've become more focused on the interior life, so that's where my pessimism comes from. I still have the wish that people would spend more time focusing on that rather than on what machines and technology can do for them.

15.

## *Information Technology*
### *(Carlo D'Asaro Biondo)*

Carlo D'Asaro Biondo is the President of Google for Europe, Middle East and Africa (and although not 'officially', also for Russia), and agreed to speak with me about the impact of digital technology. It would be difficult to find a person in Europe more qualified than him to address these issues, for he has worked in the field for his entire professional career.

The first of many conversations between us lasted more than four hours and covered a wide range of topics. Since then, the two of us have pursued many of these themes in profound ways. Carlo was the primary inspiration for the summit on the transmission of values in the digital age held at our Pontifical Lateran University (www.core-values.org), which was a significant success. After reading Carlo's interview, it will become clear to the reader why I felt that ending this book with his contribution was so appropriate.

* * *

*Who is Carlo D'Asaro Biondo? What are the stages of his formation?*

Let me give you a brief autobiography. Being born of an Italian father and a French mother, I lived in Rome until my late twenties, before attending the French school and the Department of Economics at La Sapienza University. I started my career in business consulting. From the beginning, it was clear to me that being fluent in both English and French was necessary in order to find employment internationally, so when I had the opportunity to go to Paris as General Manager and re-found the branch office of the consulting firm I worked for, I took the chance. This happened in

November of 1998. In those years, I discovered the internet and I knew that it would change the way we communicate and do business, allowing citizens of the world to have access to real-time information and communicate through a common language.

Having to start from scratch with the new consulting company in France, I decided to focus on the opportunities that the web could offer to the company for which I was a consultant: from improved terms of sale and distribution of products, to collaboration within the company. Since that time, interest in the digital world has never left me. After KPMG, I worked for some years for Unisys, a company active in the sale of software. Then, in 2004, when the AOL/Time Warner group asked me to redefine the AOL strategy in Europe and later to become General Manager of the French subsidiary and finally President for Europe, I accepted the job because I thought I could thus better understand the impact that digital technology was having on the European economic system. After that experience, I entered the Lagardère Group, with the role of International Director General. Finally, in 2009, I started working for Google, where currently I hold the position of President for Europe, Middle East and Africa, and I deal with strategic relationships, whether they be economic partnerships or relations with institutions. This is a new role for me, because until last year at Google I was responsible for the commercial activities in Southern and Eastern Europe, Middle East and Africa.

*For years I have been following the field of technology and computer science. I remember in the 1990s when Windows 3.1 was released, I was a seminarian here in Rome and I had the task of helping the seminary with computers: those were the first networks. I saw how this world developed. I agree with Eric Schmidt, who identifies the first computer revolution with the introduction of Microsoft, that is, with the PC. Now we are living the 'digital age', the 'new' revolution.*

You can identify many stages in the evolution of the internet. The internet, it should be remembered, is a protocol, that is, a language that enables computers to communicate with each other. Therefore, the connection between computers was the first step. Thanks to the internet, connecting computers to each other allows us to share or search for information in the World Wide Web. We saw the emergence of search engines in the late 1990s with the development of online search that allows users to find what they are looking for, within the vast expanse of the World Wide Web. Later, we saw the birth of social networks, through which each of us can connect with anyone, in a logic of 'one to infinity'. Now, thanks to increasingly massive mobile technology (smartphones), we see the connection between smartphones and laptops, or watches, but also in terms of home automation and much more. There comes a point when connectivity is continuous and constant, and then the internet becomes like water: water is a component of everything, just like oxygen. The internet marks such a change.

This change, especially one as profound and pervasive as the one we are living, generates fear: such a reaction is perfectly natural. I do not think we should be afraid, but I think it is necessary to accompany this process of change with a sense of responsibility. On one side of the web it has led to the radical change of business models that before worked perfectly, and for a long time. In this case, the change can be a concern for those who have to renew their models to adapt to the present and meet the challenges of the future. The examples are countless, from the transportation of persons, to universal access to information. I would like to dwell on the latter.

In the past, it was only possible to know the facts of the world by using traditional channels of communication, reading the papers or watching the news; but today, information can be transmitted by anyone, and in real time, by those who are experiencing first-hand certain events (for example, through platforms like YouTube or Twitter). Does this change, or to be more precise, this 'disintermediation' of information, eliminate the role of traditional printed media? I think not, but I think that the traditional printed media must rethink themselves, and analyse the manner in which they interact with their readers and their sources. It is not a simple process for those who live it every day, but it is inevitable and even

Google is working to deepen this change and be part of it, on the side of publishers. I am referring to the so-called Digital News Initiative, a partnership between Google and many of the major European publishers aimed at supporting quality journalism, through technology and innovation, with the purpose of encouraging the growth of a more sustainable digital news system and to promote digital innovation through collaboration and dialogue between the press and the technology sector.[1]

I gave the example of the press because I think it represents a virtuous example where there is no longer fear, but awareness and the desire to work together to figure out towards which direction to go. All companies have a reason to use the internet, perhaps in a different way: those who produce, those looking to sell, those who manage relationships with customers, those who create services. Today, we can imagine a passenger car not connected to the internet. In five years' time, you will not see any car not connected to the internet and automakers are already equipped for this scenario: there's the GPS, there's your map service, security, videos for insurance, which will cost less if the car is well equipped with these devices. You will share the car with others and hopefully it will no longer be idle 90 per cent of the time, with a significant advantage in economic and social terms. The world around us will become increasingly connected [the 'Internet of Things']. Expanding the series of examples, upon returning home your refrigerator will remind you what you need and maybe even order it for you. After all, it is simple: at any time of the day, being connected will offer some value or contribution. These moments will increase hand in hand with the imagination and creativity of those who create the services. According to Moore's law (with which you are familiar), every two years the cost of storage of information, and therefore the cost for holding information, is halved and the capacity of calculation doubles. Among other things, the time interval of two years tends to decrease with the progress of technological development. This leads to the creation of new companies, the so-called 'start-ups', that can come into existence and grow much faster than in the past. Do not forget that Google was born eighteen years ago and that entities like Snapchat, Uber and others are not even old enough to attend primary school!

*What are your thoughts on the 'sharing economy'?*

I find the development of the collaborative economy to be very interesting, leading to profound changes in the way we develop our society and how people will earn a living. First, allow me to state a fact; and second, some thoughts on the collaborative economy. The fact is that the growth of mobile telephony means that each of us is always connected to others and to the internet, and that services can be created that reach everyone, precisely because they are more useful and personalized. Example: if I have my heart rate measured and diet analysed, then an app will be able to offer health-related services at the right time, such as telling me: 'Be careful, your heart rate is increasing yet you are not under stress; go see a doctor'; or: 'Your weight is increasing, and you are poorly metabolizing some food.' Every economic sector is aware that the applications for portable devices become a means to dialogue with their own clients. Consequence: the borders separating industrial sectors come down, and new services are possible for everyone. New businesses can get into areas previously impossible for them.

Having said that, in order to understand the causes we can now look at the characteristics at the base of a collaborative economy: the full exploitation of resources, the existence of a habilitating platform, assets belonging to people (and not to businesses) that generate value, the collaborative process and, finally, digital technology as a necessary support. If we transfer these principles from the business world to the social or cultural world, we will see that collaborative innovation can lead to very important results. An example is the collaborative innovation in the social sphere, for example, after a major natural disaster. People are able to track down missing persons or give food and shelter to those who need them by sharing information and resources through digital platforms.

*What is the other aspect to which you referred?*

The second aspect is that of values and ethics to be applied to the internet. How do we make sure that all those who are part of the internet ecosystem respect values such as human dignity, tolerance and diversity? It is a matter, obviously, that does not arise only in the online world but also in the offline world, which is real life. Something that at times frightens us is that in the internet world, it seems that there are no rules, it's like the Wild West where anything goes. Well, it is not like that. The rules that apply in the offline world also apply in the online world and operators that work with the internet know that. Take for example Google: aside from respecting the laws, we have adopted a series of rules and procedures to make sure that within the products and services we provide there is nothing that sparks violence, or can be harmful to minors, or violates the copyright of the artists. Speaking of Google, there are also ethical principles about which the founders feel strongly and are therefore particularly present in the company, such as respect for diversity.[2] Google is doing a lot to promote a web that includes all, without discrimination of sex, race, religion or sexual orientation, 'the web for everyone'. To do so, we are working on four areas: 1) improving the hiring process to become more and more inclusive; 2) ensuring that Google remains a workplace in which all Googlers [Google employees] respect the rules of behaviour that take into account the sensitivity of everyone; 3) bringing computer science to groups of people who normally would not have access; and finally, 4) ensuring that all communities have access to the internet.

*When people say that internet companies have no ethics, would you disagree?*

We start from a premise. Internet companies are technology companies. These companies have their focus in the activity carried out by engineers working to create products and services that can be useful for users. The technology is neither good nor bad, the technology is neutral. It's how you use the technology, for what purposes, and in accordance with such principles that matters. Then, to be clear, user trust is fundamental: we know that if we act badly with their data or their values, we lose customers.

*In my book* Unknown Future, *my thesis is that our use of technology is the problem, not the technology itself.*

The challenge inherent in the technology, and in your question, is asking today what education must we offer to students so that they utilize the internet the best way possible? How to prevent them going off and looking for pornography or violence? How to process an education suited to our times and to create in them a consciousness, an awareness of risks and opportunities that are present online? It is a challenge for parents and educators, but also for the companies that offer online services.

*The question is: how do we introduce ethical values through the internet?*

We should do much more in order to communicate our values in the network and to talk to young people using a language that is appropriate for them. Here in France it is sometimes difficult to educate children according to the values of the Catholic Church. How can we get them to not search for specific topics on the internet? To have appropriate behaviour, which reflects moral values? We must find ways to do so. I do not have all the answers. If I had the answers I would have already solved many problems. Through the network, an ethical dimension must be present. In short, we want to launch many ideas and think about this: how can we convey these values through technology? I have the impression that today we must start from people, analysing the life of everyday interactions between people and then create, I think, groups that share the same values.

*Going back to Google services, Google+ was born in response to Facebook, or does it have another purpose?*

Google+ and Facebook are not the same thing. Google+ offers a number of services linked to people's profile, putting them in contact with others, particularly in a controlled and safe way. Unlike Facebook, you work with people you know, so your data is protected. The philosophy of Google Drive is this: know your data and decide what to do. Maybe Google did not immediately understand the importance of the social networks, yet we were among the first to understand the importance of *mobile*. In the world of the internet, sometimes you miss opportunities.

*Elon Musk, CEO and CTO of SpaceX and friend of Larry Page [co-founder of Google], said recently that he is afraid of an army of drones built by Google.*[3] *What do you think?*

Elon is quite a character. He has done exceptional things; for example, co-founding PayPal, Tesla cars and the SpaceX space agency, and now he is launching a transport system between Los Angeles and San Francisco through a high-speed tube. He is extremely creative! We must understand why he said that, if he really is afraid or if it was just a media stunt. If he thinks that such an army could be useful to get products to consumers (as Jeff Bezos plans to do with products sold on Amazon), then I can respond by saying that we are still far away. Consider, however, the importance of this project. A few years ago, I spoke with Sergey Brin [Google co-founder with Larry Page], and he invited me to think about how to drop a package from a plane without it breaking on impact. At the time I did not understand why Sergey Brin had thought of this. It seemed a strange thing to do. Then, after thinking about it, it became apparent that if we solve this problem, we solve the problem of world hunger . . . The strength of these great people in today's economy is that they think differently, they go to the root of what technology can do to solve humanity's problems. And they think (correctly, in my opinion) that if they solve a problem of humanity, it will turn out in their favour.

*Could you explain this idea a little more?*

If you could deliver what you want quickly, anywhere in the world, it would decrease the cost of moving goods and services, providing something more in the world. You would have solved a big problem! And this is not just a commercial issue but also a humanitarian issue: just think how easy it would be to provide for the most impoverished areas of the world or those hit by natural disasters.

This is not to say that Google is an NGO, but it is true that Google is constantly looking for ways to improve the world we live in.

*Please excuse my cynicism, but I believe that the problem is not a technical one, but one of wanting to solve it. I'm not sure how many people really want to eradicate poverty or help those affected by the earthquake in Nepal.*

One of the things I like about Google is that I work with people who also have a philanthropic vision. If you speak of individual countries, that they are not very interested in the rest of the world, I agree; one of the things that I like about my company is that I have the perception that we all share this desire: we really want to do good. Whether we actually achieve it or not, I'm not the only one who can say, but the firm has the desire to help improve our world. I'm not saying we do not pursue our own interests (because the more you earn the more you can help, the more problems you can solve), but I think there is this strong ethical dimension. I feel much better when I go home at night now, compared to the past, when I worked in other companies, because I can say: 'Well, we have solved some problems, we have found useful things.' Ours is a company that can afford to invest time and resources to solve problems. The company helps others even for cultural purposes. We donate money to charity: the company has an internal programme, in which for every dollar an employee gives to charity, the company doubles it. It is a company that has ethics: there will be many flaws but, for those who work there, this has always been a common experience. I feel good because there are people with valid and true ideals.

*Speaking about people's fear of machines, many articles that discuss this theme mention Google, which has become a favourite target of alarmists.*

Yes, of course, because it is easier to strike a very large and visible target [like Google]. Anyway, I do not think that being criticized is a problem; the issue is whether such criticism serves only to block technology's evolution and progress, and protect some established positions of income.

*The curious thing is that Google is a free service.*

Yes, that is why it is used by so many people.

*I am interested in the vision that Google's leaders have because, in my opinion, what will take place within the next five to ten years will depend on what Google thinks now.*

No, I do not think so. Google does not have a defined long-term vision: five years is a very long period. Google starts from a principle: there is the consumer, one experiences and listens to the consumer and it is he who tells you where you need to go. Sure, we know and we believe that *mobile* is important, so we invest in this area. With Alphabet today we try to work with so many different industries to bring them that 'web component', which becomes an ingredient to provide products and services in many sectors, from banking to healthcare, from insurance to transportation. We have some ideas, we think about some things, but we have no detailed planning; we respond to the demands of the consumer. We are constantly being surprised by things. When we developed Google Glass, for example, we did not know all the possible uses the product would have. When we launch a product, we do not know what will be the actual usage.

Trying to anticipate the consumer and guess what the consumer will want in ten years is impossible. We can identify some trends, these are clear. For example, cellular phones will not decrease, but they will continue to develop. The internet is becoming more and more pervasive and increasingly will include more users. The internet becomes a little like water or electricity: you cannot live without it, or you will be worse off without it.

Thinking beyond that is very complicated. No one can say what will happen; we do not know how the world will evolve. There are new companies, moving into new services, applying new ideas. The beauty is precisely that. I understand that for a priest this could be a problem, but reasoning it out, it's not really problematic. When we use the term religion, we usually mean a belief in God; and although God has His own vision of the established world, He has given people free will, and therefore has a vision of the world that evolves, one thing leading to another perhaps without any predictions.

*God has a vision for humanity, but we must take human freedom into account. Do you agree?*

We agree on human freedom. So why should Google impose its own vision on people if even God refuses to do so? That would be absurd! It would not make sense. It's just technology, nothing more. I live in a company that believes in man, which respects diversity, respects the way of thinking of others and which deep down possesses this aspiration: solve problems. Wanting to replace God (because, after all, this is what your question is implying) seems absurd! I do not believe it. The day on which I perceive this kind of plan, I will leave Google, and, like me, I think the majority of employees. We are 'obsessed' with all that is human, we are all geared towards this dimension.

*You have stated well, 'all things human' and not 'all things of consumers'. At the heart of research, there is not only those who use the product, but the person as a human being.*

There are many aspects to human beings: there is that of the consumer, but there is also so much more! You cannot stop only at the consumer: the reason why products like Google work is that we are interested in the human aspect. I repeat 'human' because they respond to the interests of the person. Google responds to a variety of needs, such as buying, informing, studying, playing or entertaining. These are free choices. First, there is no specified project about the future. Second, it makes no sense to have it. Third, it would be annoying to us. The fascinating thing about the work we do (and we have the means to carry it out) is the ambition to solve problems. There is the idea to create a cultural institution or spread culture through the Web: Google is interested in this. We try to respond to various ambitions. Currently, I'm interested in how to transmit moral values via the internet. I am glad that we have relations with the Church and with other religions.

*Google is working with the Vatican: I believe that there is an initiative for the digitization of the Vatican Library. Is this true?*

We have launched a number of projects, both for culture and for the spread of values, with Cardinal Gianfranco Ravasi, President of the Pontifical Council for Culture. We have devised valuable and useful initiatives to spread the values of the Church and to place some books on the network in order to contribute to the development of the Church's websites. Our leaders also want to fully understand the Pope's vision of how the world evolves, the Pope's vision of the future of the world, on what are the values and ways of life to make this world better.

*Let me provoke you again. Google offers a service and does not ask anything in return. Is it really like that?*

It is obvious that we profit from what we offer. Google's business model is based mainly on advertising, like many other media, such as television (except in cases where you have to pay a fee or sign up for a service) or radio. We have an indirect model, as is the case for many other media. It is not a new thing, this has been going on for a long time. It is interesting to identify what is behind it, because it is clear that a product which is free also becomes a huge responsibility. When our product is used by hundreds of millions of people, it must have certain characteristics, such as reliability, confidentiality and the care of privacy. With particular reference to this last point, Google places maximum attention on the development of products that protect the security and privacy of citizens by allowing them to be protected from cyber attacks (such as spam, phishing, etc.), to know how their information is being used (transparency) and to decide precisely how such data should be managed.

If we did not act in this way, no one in the world would put their data in our hands, and Google would no longer function. The fact that users feel vulnerable means that Google should be more and more attentive. The ethical code is essential. As a global company, the ethical discourse is complicated and articulated because the way we perceive and define the values of life in society is different across the globe.

*Returning to the theme of values, what problem does technology pose for us today?*

If we go back to technology, the problem this poses today concerns substance over form. How to transmit values through these tools? How do you grow as a human being? How to share the fact that today we can speak with an African, an Asian, an American, and understand each other? In two or three years we will have simultaneous translation for almost all languages of the world. We are beginning to ask new and fundamental questions. When you communicate with someone from a different reality than your own, you need to understand not only the language but also the culture which forms the basis. I often wonder where our cultures are compatible. As I travel, I make a special effort to dialogue with everyone. This is vital, because we need to discuss these topics in order to find the proper balance. At Google, because we are global, we have global principles that guide us when making choices. We try to improve every day. There is a team of people that focuses on these topics: it is a group of experts who think about these issues, for example, which customers to accept and which to not accept in advertising. We have found that there are some customers who create models designed to take advantage of our system.

*What do you mean?*

For example, buying keywords from Google, paying very little, and then creating a page for advertising that provides higher revenues, though not providing any content to users. We forbid this and we close the page.

*Is it an automatic process, a control mechanism?*

We try to have automatic mechanisms, but there are also other means of control. Automatic processes are always imperfect. Let me give you an example: the protection of Intellectual Ownership Rights. We have developed a system of protection of authors' right on YouTube called 'Content ID' that allows right-holders to protect their work in an automatic fashion. The system works perfectly and is used by the vast majority of content owners who post videos, but if some illegal content escapes the system's control, there is the possibility to report it through a procedure called 'Notice & Take Down'. We must have controls, we must have an ethic, even if imperfect. It is the question we are constantly asking, and that's why I wanted to dialogue with you. Not only to get to know you, but also because it is important for me to debate these issues. I am not in charge of Google, I am merely one of the directors, but I have a duty to contribute to everyone walking down this path. I can assure you that this company policy is vitally important for Google because of the influence that we have and the way we respond to the questions of young and old alike. We are constantly asking ourselves, 'Are we proceeding in an ethical way or not?' Every day, constantly, this is a kind of mantra.

*The world press never seems to bring out these moral dilemmas.*

Just to be clear, I did not say that we always do everything well. I said we try to! Some things we do well, others perhaps less so. Regardless, the ethical motivation is part of the company in a strong way. Take, for example, the construction of the new campus, the Googleplex. Why are we building this? Because at the Plex employees will be happier when they work. Here is another aspect not to be overlooked: the fact that our directors' salaries are mostly variable and not fixed is a very important aspect of our mentality which is geared towards results, and therefore embraces the sense of responsibility. Tomorrow, for example, the company could decide to pay me only the fixed part of my salary, if I behaved badly. Google can do this. There is also another possibility: the more you grow in the company, the greater is the variable part of your salary, because it means that you share this desire to help create something valuable. The key issue is the motivation. Then you have to accept the risk. It is true that they treat us very well, we are paid very well, we cannot complain about anything. However, the fact that there is a variable part and that there is a risk involved is fundamental to the firm. Sometimes I do not sleep at night and say to myself, 'Maybe tomorrow it will all end, and then what will I do?' Nothing is guaranteed, nothing is 'given' definitely. It is the opposite of the concept of aristocracy by birth or entitlement. We must strive every day to continue forward. This principle is deeply entrenched. It might seem strange, but our managers stay very long because the concept of risk is fascinating and challenging, requiring a certain behaviour. Besides being competent, we place great importance on a person's attitude. When we hire someone, we look at the knowledge that he or she has, and then we look at their attitude. For us this is perhaps more important than the rest, because knowledge can be learnt from someone and can be transmitted, while the attitude is linked more to the person. If a person has ways of thinking and acting that are not addressed to the outside world, to others, we do not hire that person because they would not be compatible with the values of inclusiveness, tolerance and sharing needed for self-criticism and the ability to adapt to a world in constant flux.

The first principle of the company's philosophy is the user: is the user happy? This is the starting point. We need to develop services to meet individual needs. First you think of a service and then, later, you figure out how to make a profit. When we do a business plan, the first question we ask is what problem we are solving, not how much money are we going to make. If we solve a problem, we will find ways to make a profit. Think of Google maps: we had no idea how to make money with the maps, but we said: 'These maps are useful because people will know where to go and will find what is useful, living their lives in a better way having access to information. Let's do it.' Our mission is precisely this: allow everyone to have access to information. Google's mission is to provide information to all.

Let me take the example of Gmail. All these products are made with this principle: first we create, then we try to make money from them because we have solved a problem. Our technology allows people to solve problems and thus it is useful. For this reason, we enter into many sectors, because our technologies try to solve different problems. The mobile phone in my pocket helps me solve many problems every day and can even read different parameters related to my health. It is difficult to determine how to do it correctly, but we are still trying to work to solve problems.

*Is there a difference between the impact of technology on a young person and an adult?*

I think so. That is why we invest a lot of energy in education for the correct use of digital technology. If an educational centre calls me, I always try to go in order to convey to young people what we live in the company, so that they know the work we believe in. I live this aspect personally all the time, even though with time I have changed, for as a young man I worked two or three days without sleep, but now I cannot do that any more.

*We should continue to discuss the foundations of what we believe in. If we do not ask ourselves these questions, if we do not have conversations like the one this evening, how can we improve?*

We do not have all the answers. We look for them. For example, we ask whether it is right or not to use technology to improve the condition of everyday life. The immediate answer is yes, that's right. Perfect, so we make people live longer. We should think, for example, about what happens in Europe with all the immigrants arriving. We all agree that our duty is to do something so that these people can live better lives. Because if you are born in Africa or Syria, do you have less right to live? That would be unacceptable, it is racism. Why should the concept of merit not be universally accepted? What is the role of the state over the individual? Is it I that should decide, or should the state decide for me? If the state decides for me (as is the case today in some areas), what do I do? If I believe in the state, the state must have a moral behaviour, so I recognize it and accept that it acts that way. But when I see some states and the way they act, I have enormous doubts. I believe that the individual should have the right to choose their own moral values and that the state today should not replace citizens with this incorrect and amoral system. The state does this, and does it without a reference to what is right.

*I must say that this dialogue has been a revelation for me because of the richness and the vast extent of the topics we have covered. We've covered much ground!*

These are the problems we focus on every day because we live them. What we do would not make sense otherwise. I assure you that we pose these questions all the time, but I am unsure exactly where we are headed. Our task is to invite others to dialogue with us. Every day we work on this aspect.

# *Notes*

## *Introduction*

1 See https://fcw.com/Articles/2014/02/13/How-VA-is-driving-telemedi cine.aspx. The article shows that the use of this type of technology is increasing every year: 'the reach of VA's telehealth services is growing by about 22 per cent a year.' This trend was also identified by Dr Margaret Wierman in her interview for my previous book in Italian, *Futuro ignoto. Conversazioni sulla nuova era digitale*, IF Press, 2014, 255–76.

2 See http://www.nytimes.com/2014/11/12/science/weapons-directed-by-robots-not-humans-raise-ethical-questions.html.

3 See http://www.unog.ch/80256EE600585943/(httpPages)/6CE049BE 22EC75A2C1257C8D00513E26?OpenDocument.

4 The famous demo of the algorithm 'learning' how to play Atari can be seen at https://www.youtube.com/watch?v=V1eYniJoRnk.

5 See www.particle.io for some very inexpensive devices which can be deployed in the environment and linked together in order to provide real-time information concerning a host of indicators (for a cost of about US $2 per month). With such technology, the 'Internet of Things' is quickly becoming a reality.

6 Piero Scaruffi, *Intelligence is Not Artificial: Why the Singularity is Not Coming Any Time Soon and Other Meditations on the Post-Human Condition and the Future of Intelligence*, Amazon Kindle edition, 2015, 59.

7 Steve Omohundro, 'Autonomous technology and the greater human good', *Journal of Experimental & Theoretical Artificial Intelligence*, 2014, vol. 26, No. 3, 303–315 [DOI: 10.1080/0952813X.2014.895111], 1.

8 See http://www.yudkowsky.net/singularity/aibox/ where Eliezer explains the rules of the experiment, and shows that in the majority of the scenarios, the AGI does escape.

9 Stuart Armstrong, *Smarter Than Us: The Rise of Machine Intelligence*, MIRI, 2014, 1–4.

10 One of the best books on the subject was written by two MIT professors, Erik Brynjolfsson and Andrew McAfee, *The Second Machine Age: Work, Progress, and Prosperity in a Time of Brilliant Technologies*, W. W. Norton & Company, New York, 2014. Perhaps the best (and most often quoted) study of the issue was made by two scholars from Oxford, Carl Benedikt Frey and Michael A. Osborne, with their hugely influential paper, 'The Future of Employment: How Susceptible are Jobs to Computerisation?', 17 September 2013; see http://www.oxfordmartin.ox.ac.uk/downloads/academic/The_Future_of_Employment.pdf. Although estimates vary, 40 per cent of all jobs in industrial nations are at risk of being taken over by computerization within the next twenty years.

11 Alan Turing, 'Computing Machinery and Intelligence', in *Mind* (1950) LIX: 236, 433–60 [DOI: 10.1093/mind/LIX.236.433], 443.

## *1. Man/Machine Relationship (Francesco Cassanelli)*

1 The Linate disaster was an airline accident that occurred on 8 October 2001 at 8.10 a.m. at the Milano–Linate airport, with a final tally of 118 victims; it was the most serious airline catastrophe ever to occur in Italy. The accident involved a private Cessna Citation CJ2, which had mistakenly entered the main takeoff runway of the Milanese airport and was hit by a McDonnell Douglas MD-87 of Scandinavian Airlines, which was taking off. The impact killed the occupants of the Cessna immediately and damaged the MD-87 to the point that it couldn't complete the takeoff, causing it to crash into a building used for sorting baggage, located at the end of the runway.

2 See http://www.wired.co.uk/news/archive/2014-01/15/cold-fusion-moves-into-mainstream.

## *3. The Future of Design (Don Norman)*

1 See http://www.jnd.org/dn.mss/automatic_cars_or_di.html.

## 4. *Military (Elliot Rosner)*

1 Among multiple videos about the Training Center, see https://www.youtube.com/watch?v=Yu3Df1dXCek.

## 6. *Advertising (Sir Martin Sorrell)*

1 See the story of the controversy here: http://www.bloomberg.com/news/articles/2014-02-14/harvard-professor-attacking-google-thrives-as-web-sheriff. It took place in February 2014.
2 See the story about banning advertisements in UK cinemas which included the Lord's Prayer here: http://www.theguardian.com/film/filmblog/2015/nov/24/lords-prayer-row-can-cinemas-really-ban-adverts-church-of-england-dcm.
3 Read the complete interview here: http://www.mirror.co.uk/news/business/terror-attacks-threaten-world-economy-6894117.

## 8. *Cyber Security (Dave Aitel)*

1 See the FBI Director James Comey's testimony before Congress: https://www.districtsentinel.com/fbi-director-continues-crusade-against-encryption-calls-on-congress-to-act/.

## 9. *Journalism (Christopher Altieri)*

1 The Pontifical Gregorian University is run by the Jesuits and caters primarily to the formation of priests and those studying for the priesthood. At the same time, there are many lay students who study there.
2 Christopher Altieri, *The Soul of a Nation: America as a Tradition of Inquiry and Nationhood*, Pickwick Publications, 2015.
3 Marshall McLuhan, *Understanding Media: The Extensions of Man*, Mentor Publishing House, New York, 1964. Later, he published the book

with the title *The Medium is the Massage: An Inventory of Effects* in 1967 with Bantam Books. Reportedly, the title was a typo by the printer ('Massage' instead of 'Message'), but McLuhan left it as it was.

4 The event occurred on 5 September 2015 in the early evening. See http://www.breitbart.com/big-government/2015/09/05/pope-francis-sneaks-out-of-vatican-to-have-eyeglasses-repaired/.

5 See http://www.ibtimes.co.in/did-isis-destroy-fake-palmyra-statues-monitoring-group-says-militant-group-plans-smuggle-638093.

6 American politician Anthony Weiner, former member of the United States House of Representatives from New York City, has been involved in two sexual scandals related to sexting, or sending explicit sexual material by cell phone. The first, sometimes dubbed Weinergate, led to his resignation as a congressman in 2011. The second, during his attempt to return to politics as candidate for mayor of New York City, involved three women Weiner admitted having sexted after further explicit pictures were published in July 2013. See Wikipedia.org.

## 11. *Existential Risk (Anders Sandberg)*

1 See http://futureoflife.org/2015/10/12/ai-safety-conference-in-puerto-rico/.

2 See the announcement here: http://futureoflife.org/2015/10/12/11m–ai–safety–research–program–launched/.

3 See http://www.oxfordmartin.ox.ac.uk/downloads/academic/The_Future_of_Employment.pdf.

4 See http://www.nobelprize.org/nobel_prizes/chemistry/laureates/1980/berg–article.html. From the article by Paul Berg, Nobel Laureate in Chemistry (1980), 'The lofty status of the Asilomar Conference and the deliberative process it spawned stems from its success in identifying, evaluating and ultimately mitigating the perceived risks of recombinant DNA. Looking back now, this unique conference marked the beginning of an exceptional era for science and for the public discussion of science policy.'

5 See Yudkowsky's paper with Nick Bostrom: https://intelligence.org/files/EthicsofAI.pdf.

## *12. Philosophy (Johan Seibers)*

1 The precise reference is the following: 'If you went in search of it, you would not find the boundaries of the soul, though you traveled every road – so deep is its measure [logos].' Diels-Kranz, Collection of Presocratic Sources, 22B45.

2 The precise book referred to is the following: C. K. Ogden and I. A. Richards, *The Meaning of Meaning: The Study of the Influence of Language upon Thought and of the Science of Symbolism*, Harcourt Brace Jovanovich Publishers, 1923.

3 The speech referred to is from an address on 22 October 1996 to the Pontifical Academy of Sciences. The original French text can be found here: http://w2.vatican.va/content/john-paul-ii/fr/messages/pont_messages/1996/documents/hf_jp-ii_mes_19961022_evoluzione.html.

4 Nick Bostrom, *Superintelligence: Paths, Dangers, Strategies*, OUP, 2014.

5 This refers to the 27 June 2015 terrorist attack on the city of Sousse in Tunisia, in which thirty-nine people (mostly foreigners) were killed: http://www.bbc.com/news/world-africa-33287978.

## *13. Nuclear Instruments (Troy Anderson).*

1 See recent developments at Google along these lines: https://www.theinformation.com/with-apple-in-mind-google-seeks-android-chip-partners.

2 See https://www.washingtonpost.com/news/the-switch/wp/2013/10/21/yes-terrorists-could-have-hacked-dick-cheneys-heart/.

## *14. Mechanical Engineering (Bill Shores)*

1 The element with sixty-six electrons is called Dysprosium and its classification is a 'Rare Earth' element.

2 The project was actually sold to private investors who kept the satellites functioning. It is now called Iridium Communications Inc., with central headquarters located in McLean, Virginia. They give service to the United States Department of Defense, as well as maritime and aviation businesses.

## *15. Information Technology (Carlo D'Asaro Biondo)*

1 See https://www.digitalnewsinitiative.com/.

2 As an example, the directors of Google have established an Ethics Council, which meets periodically to address ethical issues surfacing in the digital world. See https://www.google.com/advisorycouncil/.

3 See the article by Andrew Griffin in the *Independent*, 13 May 2015. Musk asserts that the development of drones with artificial intelligence might have unintended consequences, representing a threat to humanity.

# *Acknowledgements*

First and foremost a word of gratitude to the people who were kind enough to allow me to interview them for this book. Without them, their time and their insights, *Connected World* would not have been possible. A colleague of mine once told me, 'If you want something done, ask a busy person.' The rationale behind this proverb is that if a person is busy, it is because he or she gets things done. All of the people interviewed in the book qualify for that title, and I am very thankful for their valuable contribution.

A very special word of thanks goes to Carlo D'Asaro Biondo, not only for sponsoring the book, but also for believing in the project in the first place. After meeting with him for the first time in Rome, my life changed in a significant way and together we have been able to achieve many things. *Connected World* is just another step in the right direction.

A very dear friend, Jorge Rodríguez, told me four years ago that I was wasting my time doing research in this area. Throughout our many conversations, I tried to persuade him of the relevance of this field. Thanks to his scepticism and my desire to prove him wrong, I kept plugging away and managed to get *Unknown Future* published (in Italian). He has since been named Auxiliary Bishop of the Catholic diocese of Denver, Colorado, continues to be my friend, and is glad that I did not take his advice.

Also four years ago, another good friend, Stuart Fulton, Senior Partner at PricewaterhouseCoopers and Chief Officer for Information, after a series of long conversations, told me: 'You should write a book.' At the time, I just shook my head and said, 'Yeah, sounds like a good idea.' Thanks, Stuart!

Stefano Forte, a former philosophy student of mine at the Pontifical Lateran University, spent many hours labouring over the three interviews in Italian. His linguistic skills and logical rigour helped to transform raw dialogue into coherent copy. Stefano believed in the project from the very beginning, even though at the time it seemed like a pipe dream.

Peter Waymel offered expert help translating into English. One of the interviewees even remarked that he preferred the English translation to his original Italian. More often than not, translators tend to ruin original texts, but Peter did an excellent job of bringing out the best of the texts into English.

The great people at Penguin deserve a lot of credit. Fred Baty, especially, spent long hours poring over initial drafts, and gently yet clearly suggested well-needed improvements. Karen Whitlock added a professional touch with her editing. Although I procrastinated for long stretches, we made our deadlines and completed the book on schedule.

Finally, I am very grateful to all my philosophy students at the Pontifical Lateran University who had to put up with my frequent tangents concerning the digital world in class, and supported me along the journey. They were a constant source of motivation and valuable feedback.